McCORMICK
FARMALL
A

D1208584

FARMALL

The Red Tractor that Revolutionized Farming

Randy Leffingwell

Voyageur Press

Dedication

For Arden, Bob, Guy, Jay, Jerry, John, Keith, Ken, Louis, Martin, Mike, Neal, Wayne, and Wilson. I hope you enjoy "our" book!

First published in 2005 by MBI Publishing Company LLC. This edition published in 2007 by Voyageur Press, an imprint of MBI Publishing Company, Galtier Plaza, Suite 200, 380 Jackson Street, St. Paul, MN 55101-3885 USA

ISBN-13: 978-0-7603-3047-0
ISBN-10: 0-7603-3047-6

Editor: Peter Schletty
Designer: LeAnn Kuhlmann

Printed in China

Credits:

On the front cover: Farmall Model 400 high-clearance diesel, owned by Bob and Mary Pollock.

On the frontispiece: 1941 Farmall Model A, owned by Vercel and Marilyn Bovee.

On the title pages: 1956 Farmall Model 350V high-clearance, owned by Bob and Mary Pollock.

On the back cover: 1938 Farmall Model F-12. 1954 Farmall Model M-TA. 1972 Farmall Model 1468.

Contents

ACKNOWLEDGMENTS

This is the second time in my life I've been able to write about and photograph Farmall tractors. As with that first book in 1999, I must thank my friend and colleague Guy Fay for his help with this book. He practically has made the study of International Harvester Corporation, and its tractor and implement development, his life's work. His insight and understanding of how IHC and Farmall have affected the history of mechanized farming are second to none.

I wish to thank John Harper at CNH America LLC for access to information and photographs of the new Farmall model D and DX tractors. I also wish to thank Unette Lemke, CNH Communications, for permission to quote the interviews included in the final chapter of this book. I am grateful to Jeff Walsh, former CNH Director of Communications, for making available QC-503.

Many tractor makes inspire loyal legions of collectors. In 15 years of doing books on these machines, I have met owners and enthusiasts of all makes. Among the very most loyal are the hundreds of enthusiasts I met and worked with while researching and photographing the IHC grey and red tractors for this book.

A number of people opened their barns and sheds to me, washing up and pulling out a grand array of International Harvester's Farmall history.

I am deeply grateful to John and Jane Alling, Valley Center, CA; Mike, Linda and Eric Androvich, Grand Rapids, OH; Arden Baseman, Mosinee, WI; Dave and Anita Boomgarden, Chatsworth, IL; Vercel and Marilyn Bovee, Alto, MI; David and Ash Bradford, Warren, IN; Nate Byerly, Noblesville, IN; David and Gail Fay, Greenville, PA; Keith and Cheri Feldman, Alto, MI; Lyman and Vivian Feldman, Alto, MI; Frank Ferguson, Decker, MI; Bob and Michelle Findling, Gladwin, MI; Allan and Joan Fredrickson, Lakewood, CA; Wilson and Portia Gatewood, Noblesville, IN; David and Carol Garber, Goshen, IN; Jack and Tammy Gaston, Athens, OH; Jay Graber, Parker, SD; David and Linda Grandy, Waconia, MN; Rod and Becky Groenewald, Director, Antique Gas and Steam Engine Museum, Vista, CA; Dave Hinds, Marion, IN; Joan Hollenitsch, Garden Grove, CA; Ken Holmstrom, Harris, MN; Wayne and Betty Hutton, Clarence, MO; Matt Jackson, Noblesville, IN; Kenny and Charlene Kass, Dunkerton, IA; Wendell and Mary Kelch, Bethel, OH; Jeff Kelich, Arcadian, IN; Tom and Mark McKinney, Noblesville, IN; Harold McTaggart, Port Hope, MI; Jerry and Joyce Mez, Avoca, IA; Judy Meyer-Diercks, Stonefield Village, Wisconsin State Historical Society, Cassville, WI; George and Barbara Morrison, Gladwin, MI; Ron Neese, Noblesville, IN; Robert Off, Tipton, IN; Scott Parsons, Oceanside, CA; Jay Peper, Toledo, OH; Loren and Elaine Peterson, Sparta, MI; Bob and Mary Pollock, Dennison, IA; Fred and Janet Schenkel, Dryden, MI; Greg Schmitt, Noblesville, IN; Denis Schrank, Batesville, IN; Neal and Shirley Stone, Wisconsin Dells, WI; Lawrence Terhune, Princeton, NJ; Martin, Marsha and Matt Thieme, Noblesville, IN; Stew and Pat Thomet, Alto, MI; Bill Tyner, Westfield, IN; John Tysse, Crosby, ND; Denis and Linda Van de Maele, Isleton, CA; Mike and Paul Van Wormer, Frankenmuth, MJI; John and Barbara Wagner, White Pigeon, MI; Gary and Judy Walton, Imlay City, MI; Norm and Ardeth Walton, Imlay City, MI; Warren and Janice Walton, Imlay City, MI; Louis, Linda and Tim Wehrman, Reese, MI; and Bob, Kathy and Randy Zarse, Reynolds, IN.

I must thank Peter Schletty, my editor at MBI Publishing Company, for his thoughtful edits and suggestions in bringing this book into print. I want to add thanks to my friends, Darwin Holmstrom, senior editor, and Zack Miller, publisher at MBI, for asking me to revisit this ever-interesting subject, and most especially to Tim Parker, director of worldwide publishing at MBI, for launching my interest in farm tractors 15 years ago.

Lastly, I am deeply grateful to my partner in life, Carolyn, for her love and encouragement and for so much more.

— Randy Leffingwell
Santa Barbara, CA

FOREWORD

The first tractor I ever drove was a Farmall. It would probably be a safe guess that the same is true for many of you reading this book. With more than 5 million tractors produced and sold around the world, many of them still running strong today, the Farmall was perhaps the single most popular tractor line in agricultural equipment history.

But its popularity alone is not what gives Farmall a special place in history. Farmall was the tractor that allowed a tremendous number of farmers to turn in their horse teams for mechanical horsepower, helping to enable agriculture's production revolution. Farmall wasn't the first tractor. It wasn't even the first tractor from International Harvester, but it was the first tractor that was practical for all farmers.

Unlike the competition of the time, the first Farmall tractors were lightweight and nimble, with a high power-to-weight ratio. The Farmall worked in all situations with a series of matching implements for plowing, cultivation, and harvesting, as well as a reasonable price tag. Like its name, it was designed to "Farm-All." In an era when agriculture dominated the countryside, Farmall tractors became a regular sight from roads across the United States.

But what made that original Farmall—the ancestor of generations of bright red agriculture machines —so successful? Was it the superior engineering? That was part of it. Was it the progressive design that allowed one machine to multi-task for more different jobs than anything that had come before? That was a big part of it too.

Farmall represented an unwavering commitment to agriculture. Perhaps that is one of the reasons farmers identify so strongly with their tractors. Besides themselves, there was nothing on the farm that worked harder.

That commitment to agriculture has survived multiple generations of the company, and I continue to see it alive and well at Case IH today among dealers, field and customer service representatives, engineers, marketers, and customers. It's one of the many things that makes me proud to have been associated with the company.

It also helps explain why we brought back the Farmall name. Farmall provides Case IH with an opportunity to look to the future while honoring the past. Our future successes are built upon the strong foundation provided by legendary products like Farmall. That's why I am so honored to have the opportunity to introduce you to this extraordinary book. Besides the beautiful photography you'll see on every page, you will also get a sense for how the Farmall tractor helped revolutionize farming.

—James L. Irwin
Former Vice President, Case IH North American Agricultural Business

Deering and McCormick Prepare for the Future

The "Famous" engine turned a 12-inch diameter pulley. Operators moved a lever that shifted the engine to bring the pulley into contact with the 51-inch-diameter friction drive wheel.

Chapter 1

1870–1913

John Steward was irked. While many historians have recounted the history of International Harvester Corporation (IHC)—and a number of those specifically have examined its farm tractor successes and failures—Steward, as an IHC insider, was perhaps the first. At the time, he believed that Cyrus McCormick's family was revising facts to bolster its name and company reputation. Steward had worked as William Deering's superintendent for McCormick's biggest competitor, Deering Harvester Company, in business in Chicago since 1870.

Cyrus Hall McCormick moved to Chicago in 1847, a year after his father died. Cyrus wanted to be nearer to the Great Plains west of the Mississippi River than he was at home in Virginia. McCormick Harvester Company had licensees manufacturing reapers in Cincinnati, Ohio, and in Brockport, New York. The McCormick operation was a family affair; Cyrus' youngest brother, Leander, was production supervisor and the middle brother, William, served as family business manager. Years later, Leander, by then vice president of production, apparently became jealous of Cyrus' fame and claimed that their father, Robert,

not Cyrus, was the reaper's true inventor: that Robert was its builder, and that Cyrus had been merely a salesman.

Cyrus fired Leander and his son, Hall, in April 1880 as the factory geared up for harvest-season manufacture. Cyrus hired Lewis Wilkinson, who had managed production at the Colt firearms armory in Hartford, Connecticut. Wilkinson quickly established a night shift to meet equipment orders. After graduating from Princeton University, Cyrus McCormick Jr. arrived home in Chicago to find his father excited by the factory for the first time in years. Cyrus Jr. watched Wilkinson introduce precise manufacturing techniques and the concept of interchangeable parts. After Wilkinson left in April 1881, Cyrus Jr. took over. During the next few years he developed several more machine tools capable of producing large quantities of identical parts. Production that reached 17,500 implements in 1880 increased to 31,000 in 1881, then to 46,000 in 1882, and to more than 100,000 in 1889, truly reaching mass-production levels. The European "craftsman" approach of the 1870s had used skilled and talented blacksmiths and

Few people would call it a factory, yet this tent, photographed on August 13, 1910, was home and birthplace for IHC's large 45-horsepower Mogul tractors. The big twin-cylinder machines had outgrown the Akron Works, so IHC's Executive Committee moved them home to Chicago—where it also had no space. So the EC pitched a tent.

machinists to manufacture one or two products at a time. Wilkinson's "American system" used craftsmen to make patterns from which semiskilled workers produced and finished thousands of parts to be assembled, dozens at a time, by less skilled laborers.

John Steward had endured McCormick claims and boasts for years. Deering, McCormick, and others had nearly merged in 1891, but terms seemed unequal to all the companies involved. Competition intensified into the "harvester wars" that cut prices and strained every maker. In 1897, Deering, retiring after a lifetime of business success, offered to sell out to McCormick. But he couldn't agree with McCormick on values of their holdings. So the two companies, controlling nearly two-thirds of the harvesting machinery production and sales in the United States, resumed their rivalry.

One year later, in 1898, the Honorable Ferdinand W. Peck, as American Commissioner General to the Paris Exposition World's Fair of 1900, nominated Deering Harvester Company alone among all American makers of harvesting machinery "as the proper one to make the

This was John Steward's experimental tractor, at work in 1910. His clever rear-axle configuration allowed the operator and the tractor to ride level while running a lowered wheel in the freshly plowed furrow. The engine drove the front wheels by a chain. IHC records suggest they produced ten of these.

This 1912 early production Mogul 10-20 worked fresh ground behind the Tractor Works. IHC manufactured only about nine of these models in 1912, and about 75 in 1913. The long levers operated two plows each.

Cyrus McCormick learned that rival William Deering was showing a self-powered mowing machine at the 1900 Paris Exposition World's Fair. Not to be outdone, McCormick quickly assigned engineer Ed Johnston to create one for their own display.

1900 Auto Mower. Ed Johnston's machine looked simple but it represented sophisticated thinking and engineering. This machine provided its operator with a power take-off that could be disengaged if the mower bogged down in thick grass.

Johnston's first one-cylinder model wasn't strong enough when it came time for practical tests. He completed a two-cylinder version by August 1900 and in new tests, his machine won against Deering's model.

retrospective and historical exhibition." Deering asked Steward, now his patent expert, to prepare the exhibit. Steward's display comprised 17 showcases containing 96 working scale models of mowers, reapers, and automatic binders.

George H. Ellis, who joined Deering Harvester in 1889 as a 23-year-old engineer, had introduced an Automobile Mower in 1894. Ellis had begun experimenting with gasoline engines in the spring of 1891 at his home. He demonstrated his 70-pound, 6 horsepower vertical two-cylinder engine to Deering. They moved it to Steward's basement and mounted it on a Deering New Ideal mower, showing that new product to Deering and his plant superintendent B. A. Kennedy in October 1891. He began at once to develop a 16-horsepower two-cylinder opposed engine. As work progressed, Ellis mounted the same engine on a mower. Deering began to produce a few of these "Auto Mowers." Commissioner Peck encouraged Deering to include the Auto Mower at the 1900 Paris Exposition to demonstrate new technology. McCormick, uninvited, assigned his staff to create a two-showcase exhibition with five scale models and a full-size version of their own just-built, and similarly named, Auto-Mower.

Cyrus Jr. believed his company had to be one of the 1,600 exhibitors at the Paris Exposition. He had set Edward A. Johnston to work quickly. Johnston had joined McCormick's Experimental Department at age 15 in 1893. Johnston completed his first air-cooled gas engine in 1897. He mated his second version to a two-speed forward, one-speed reverse transmission in a carriage in 1898. Then he mounted one of his two-cylinder engines on a Bert Benjamin–strengthened cutter that McCormick christened the Auto-Mower (with a hyphen).

This piece of history is one of the jewels of the Wisconsin State Historical Society's collection at Stonefield Village.

Both Johnston's machine and Ellis/Steward's version were innovative, using one power source to propel the mower and power the cutters. The Ellis/Steward patent (granted in 1903) addressed the technology of transmitting engine power through gears and shafts to the functional portions of the implement in an early form of power take-off (PTO). Their system was not independent of the transmission, however, and could not be disconnected. Under heavy mowing conditions, the load on the cutter bogged the engine. Ellis' engine would run forward or backward so his transmission had no reverse gear. Getting the stalled mower out of tall grass required restarting it rotating the other way. During the test its blade jammed with a heavy growth of alfalfa. Ellis' operator could not restart the engine.

Johnston's PTO worked from his two-speed transmission that operators could disengage from forward or reverse motion. In tall or heavy grass, the operator stopped the drive wheels while the tough cutter bar chopped through the thicker crop using full engine power. The McCormick Auto-Mower outperformed all competition. Johnston applied for a patent for the machine in 1902 to protect both the reverse-gear transmission and his "independent" (live) PTO.

Neither Deering nor McCormick put these machines into full production. Each company needed the money they might use developing automotive mowers instead to fund manufacture of stationary engines and other products. McCormick, despite his competitiveness, admired

Deering's corporation and its organization. McCormick's staff were skillful salesmen, but they held large accounts receivables. Deering's people were efficient collectors and they controlled every element of manufacture. Deering owned its own steel mills. Cyrus Jr. and Deering's sons tried again in late 1901 to get together. They accepted that each company had what the other needed. One more time, mutual distrust thwarted them.

Then, in February 1902, Judge Elbert H. Gary contacted McCormick. Gary, chairman of United States Steel, sold much of his output to McCormick. Gary knew Deering's mills and foundries cut its production costs. He sensed that Cyrus Jr., hoping for similar economies, might establish his own. Gary proposed that consolidating operations with Deering might be a way for McCormick to cut costs, an idea Cyrus Jr. accepted.

There are several renditions of how the 1902 merger came about. One version suggests that Cyrus Jr. asked John D. Rockefeller to acquire Deering. Cyrus' brother Harold had married Rockefeller's daughter, Edith, in 1895. Rockefeller offered to loan them money instead, but Cyrus Jr. believed the only way to get Deering's cooperation was through outside intervention. Rockefeller recommended going to J. P. Morgan's lawyer, Francis Stetson, and in June 1902,

1908 20-Horsepower Friction Drive. Born out of Morton's Traction Trucks manufactured in Upper Sandusky, this line of gas traction engines became a successful product for IHC. The corporation installed its big single-cylinder "Famous" engines on them and sold 14 in 1906 and 153 in 1907.

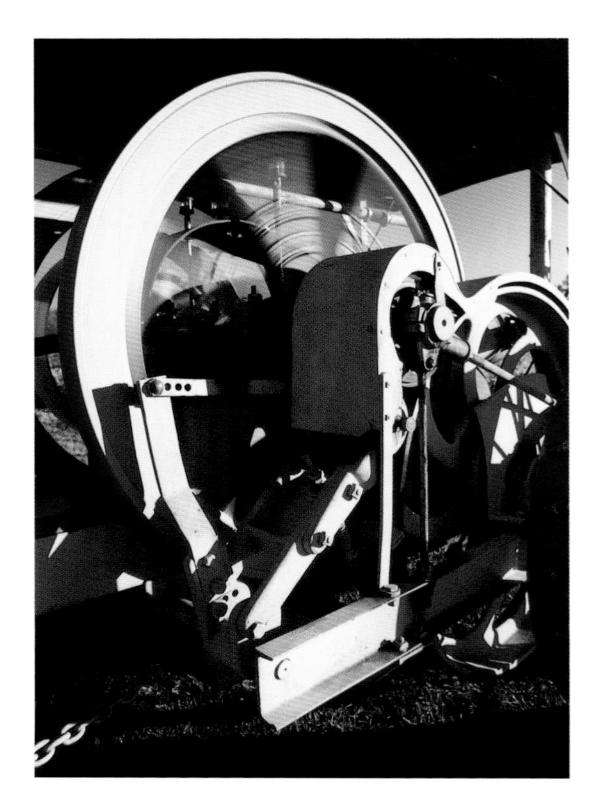

The 20-horsepower version used a single-cylinder engine with a 9-inch bore and 15-inch stroke. IHC moved the industry forward with this machine, and in 1908, it manufactured 629 in 10-, 12-, 15- and 20-horsepower variations.

Stetson contacted George W. Perkins, a partner at Morgan and Company, to arrange a meeting.

Perkins and the "House of Morgan" accommodated. Perkins not only had directed Judge Gary's U.S. Steel's merger, but Morgan also had partially financed the deal. Now Cyrus Jr. and his brother Stanley invited Perkins to engineer another merger, recognizing that they and Deering each would want to control the resulting organization. Perkins convinced the two harvester manufacturers that consolidation meant self-preservation. He avoided the matter of control simply by postponing it, meanwhile creating a voting trust where he was the tiebreaker.

On August 12, 1902, McCormick, Deering, and three other harvesting equipment makers consolidated in business under the name International Harvester Company. Deering and McCormick now controlled nearly 90 percent of grain binder production and about 80 percent of the mowers in the United States. (For its facilities and cash investment, the McCormicks received 42.6 percent and the Deerings received 34.4 percent of the IHC stock.) McCormick's growing sales force throughout Europe and beyond motivated J. P. Morgan to add the word "International" to the group's new name, implying that the conglomerate's goal was not a domestic monopoly but a worldwide investment opportunity. IHC named Cyrus McCormick Jr. its president and appointed Charles Deering, William's son, as chairman of its ruling organization that held the company stock in trust for 10 years. Morgan and Perkins capitalized the new company at $120 million (about $2.64 billion in 2005 dollars) based on $60 million in assets of its five companies, as well as $50 million in accounts receivable, and an additional $10 million in new IHC stock it sold to Morgan in exchange for operating cash. Perkins took a $3 million fee (nearly $66 million in 2005) and effectively ran the company, keeping the former rivals at peace.

Deering, McCormick, and Perkins had plenty on their minds following the creation of IHC. Deering wanted to buy every harvester maker in North America, putting an entire industry under their control. Perkins and McCormick realized this invited U.S. government scrutiny and that some makers weren't worth owning. Still, IHC picked off its competitors.

The variety of its holdings made more harvesters widely available; every farmer who needed one could get one. With harvesters in every barn, demand slowed, so IHC reduced production. Farmers began wondering about other machines to work the soil or to open untouched lands. When Deering merged its resources into IHC, the conglomerate relocated some of its raw materials refinement capability among various plants, to centralize manufacturing nearer to major distribution and shipping resources. This freed shop space to develop and manufacture new implements and machines.

The big flywheel served both to keep engine momentum going and also to govern engine speed. Production in 1908 came not only from the original Akron plant but also from IHC's facilities in Milwaukee, Wisconsin.

William Deering encouraged George Ellis to continue his gas-engine projects. In 1898, Ellis had bolted his 8-horsepower single-cylinder horizontal engines onto a sliding framework mounted on a chassis he fabricated at Deering Works. Moving the engine forward or backward made contact with friction wheels chained to the rear drive axle. McCormick and Deering jealously guarded new developments. No one shared ideas. Perkins attempted to enforce order by naming Clarence S. Funk as IHC general manager. Funk had managed the Champion firm and carried no loyalty either to McCormick or Deering. Funk's independence, however, couldn't discourage the "us-against-them" mentality between divisions.

Looking to the future, IHC purchased a Morton Traction Truck in early 1905. Several experiments led to the idea of mounting IHC's 15-horsepower Famous one-cylinder stationary engine on a framework that moved forward or back on casters. It took work to revise IHC's stationary engines to propel Morton's Traction Trucks, but the effort was successful enough that IHC ordered another 14 and moved to acquire the Akron, Ohio–based manufacturer.

Sometime in 1905, IHC's Executive Council (EC) created a product planning and review group informally called the New Work Committee (NWC); on April 20, 1906, that name became official. IHC shifted some of its binder and harvester manufacture from Milwaukee down to Chicago and relocated gas engine production from the Deering and the McCormick Works in Chicago to the nearly vacant Milwaukee plant. It established separate Engineering and Experimental Departments there and named Harold A. Waterman as plant superintendent. IHC's initial 14 buyers loved Morton's Traction Trucks, so IHC's Manufacturing Department

1911 22-45 Horsepower Mogul. This was one of IHC's first products from its Chicago Tractor Works, the facility that began life as a big tent in the winter of 1910. By 1911, the reliability and popular demand of this tractor led IHC to manufacture 583 that year.

purchased more of them for 1907, "knocked down" or unassembled, for manufacture in Milwaukee. IHC sold 153 that year.

Acquiring Keystone Company in 1904 brought to IHC not only its Rock Falls, Illinois, plant but also returned Ed Johnston to International Harvester. He had resigned in 1902 over patent ownership disputes. Now he resumed gas engine development in 1905 at Keystone Works. McCormick liked Johnston's Auto Buggy and believed it could meet farmers' hauling needs. Johnston revised it to run 20 miles per hour, carry a ton of cargo, and climb a 25 percent grade. Once IHC began to manufacture Auto Buggies in February 1907, the EC shifted Johnston and his experimental projects to a larger space at McCormick Works in Chicago. Then, in October 1907, IHC interrupted Johnston's work again, transferring him to Akron Works. There he mated his new, three-cylinder 40-horsepower gas engine to an improved version of his two-speed-forward-one-speed-reverse transmission that he had used on the Auto-Mower. He installed this into a three-wheel frame.

Expecting to repeat its Paris success in 1900, IHC entered Johnston's tricycle (and two other Morton-type friction-drive four-wheelers) in the Winnipeg Industrial Exhibition's First

Agricultural Motor Contest in Manitoba, Canada, in July 1908. The word "tractor," attributed to a sales director from competing Hart Parr in Iowa in 1906, began replacing the more cumbersome term "traction engine." Of seven gas tractors entered by five producers (five steam traction engine makers competed as well under "Light Tractor" rules for those weighing less than seven tons), Johnston finished second by less than one point. Solid performance in Winnipeg generated sales north and south of the border in areas that most needed power farming. Contest results and sales performance influenced how nearly every manufacturer planned its next machines.

IHC's Ed Johnston conceived and directed manufacture of this two-cylinder giant. Early photographs show these tractors pulling as many as 15 or 18 plows, turning over a swath of earth 20-or-more feet wide. Such plow loads sometimes broke the frames of these tractors in the early days.

Johnston was equally successful when his Auto Wagon came out. Akron Works manufactured these until 1912, when IHC renamed the vehicle the Motor Truck. Johnston, 30 years old in 1908, designed the engines that powered IHC into long-term truck success and into the tractor business. Morton's truck manufacturing methods were imprecise despite an investment in cast gears. It matched gears only to individual trucks or tractors, not for the series. They had no standard pitch or diameter. Akron and Milwaukee built to IHC precision. Akron cut gears to the nearest standard dimensions. Milwaukee Works engineers completely dismantled the tractor to develop specifications. John Steward, Ed Johnston, and his assistant, H. B. Morrow, designed a new gear train in Akron. Thereafter, IHC manufactured all its tractors from drawings rather than assembling them as pieces reproduced from samples. Johnston reinforced Morton frames, enlarged and strengthened the bull gears, and improved the cooling and brake systems.

This made IHC's gear-driven Type A tractors strong and rigid enough to pull wagons or operate threshers off their belt pulleys, but they still couldn't plow as well as IHC wanted them to. Steward examined the Harpstrite tractor produced near Decatur, Illinois. Andrew Harpstrite mounted his moldboard plows directly beneath the center of his three-wheel tractor, which featured a large drum-drive rear wheel. Henry B. Utley, IHC's purchasing manager on the New Work Committee (NWC), wanted Steward to acquire patents on every tractor and plow so IHC owned the technology to experiment or produce without infringement concerns. Steward judged many competitors "not worth having."

On June 1, 1908, Milwaukee superintendent Harry Waterman appointed young P. R. Hawthorne to design a new tractor for IHC. The market demanded ever-larger machines for first plowing to open the prairies and Great Plains. By fall, Hawthorne's 45-horsepower

It was a pretty simple operator's platform. Yet managing one of these brutes took massive arm strength. Johnston's engine featured 9.5-inch bore and 12-inch stroke and ran at about 335 rpm.

Upper right

The starting engine was one of Milwaukee Works' innovations with this big machine. One cylinder burned gas while the other compressed air. The operator pulled a lever and injected the air into the main engine to begin moving the large pistons.

two-cylinder, dual-crankshaft engine ran well. IHC introduced it as the 45-horsepower Reliance, but later renamed it the Titan. Ed Johnston worked on his own two-cylinder opposed engine that went into production as the 45-horsepower gasoline Mogul. IHC's automobile and truck plants constantly pulled Johnston away to solve their production problems, yet he completed two or three running prototypes in time for the July Winnipeg trials. A few production models followed before the EC transferred the project to Milwaukee's greater production capacity. Charles Longenecker took over the project there. Though IHC called it the Type C, NWC reports always referred to it as the "Johnston-Longenecker-Akron." It incorporated so many differences from previous efforts derived from outside makers that this machine was IHC's first complete tractor.

On December 12, 1908, IHC's president Alexander Legge assembled the NWC. Johnston and Waterman commuted to the Chicago meeting. Legge contended that farmers wanted tractors that could plow 10 acres a day with a five-bottom plow. Regional sales managers from Canada, and Eastern and Midwestern states concurred. He assigned his two engineers to create a two-forward-speed, maximum 45-horsepower two-cylinder tractor prototype as well as a 25-horsepower single. Legge continued this strategy for years, developing an intense rivalry between the two engineers and their departments that resulted in diverse solutions to engineering problems.

1911 Mogul Type C. Ed Johnston first conceived this 25-horsepower model based on earlier Akron, Ohio, developments. But other manufacturing problems pulled him off development. IHC's executive committee transferred the project to Milwaukee Works for Charles Longenecker to complete.

Canadian organizers scheduled the 1909 Winnipeg Trials for early July. On May 24, Legge approved entering eight tractors. Hawthorne and Waterman had six weeks to prepare and build the two barely finished Johnston-design machines. Unlike prior models, where bull gears drove pinion rings inside the large rear drive wheels, these used differentials and revolving axles. L. B. (Leonard) Sperry took over the Experimental Department and put workers on two- and even three-shift schedules. In 40 days, crews completed two tractors. They tested them for one single day and shipped them to Winnipeg, painting them on the moving railroad cars. Entered in a new class without weight limits, the massive machines were so heavy they both bogged down under their own bulk.

On September 22, John Steward reported to Legge that Victor, Famous, Rival, and Keystone were names not yet registered by any other maker for tractors. In August, domestic and foreign sales managers meeting with production and manufacturing leaders had proposed renaming Famous tractors after McCormick and using Deering for the Reliances; exports should be "International" for Europe and "Champion" for South America. On December 13, the committee

This tractor, moving from one factory to another like a vagabond, earned the long internal nickname of the Johnston-Longenecker-Akron tractor. It was IHC's first true tractor product, moving well beyond what McCormick's earlier efforts had accomplished.

abandoned Reliance, naming that entire line "Titan," the name in Greek mythology given the son of Uranus (heaven) and Gaea (earth) who was "of gigantic size and enormous strength." This was becoming the IHC tractor legacy: size and strength.

John Steward had new projects going at Deering Works by mid-1909. Obsessed with making tractors more useful, he built a large-scale Auto-Mower created from 16-foot grain header-binders normally pulled by horses. Johnston's Akron Experimental Department nearly had finished revisions to his opposed-twin-cylinder, Famous 45-horsepower tractor in the fall of 1909. Large tractors still were in demand in the North and West. Output of all manufacturers in 1909 reached about 1,000 gasoline tractors, 698 of which went to Canada. Johnston's products outgrew the machine tools at Akron, a plant adequate to produce Auto Buggies but not the large castings for big twins. Based on Johnston's latecomer success with his big tractor, the EC moved ahead into the tractor business, devising a solution to Akron's crowding. They agreed to construct a tractor assembly plant next to the McCormick Works in Chicago. Cyrus Jr. transferred Johnston there, naming him overall superintendent.

When he arrived on January 30, 1910, there was no building yet and few tools with which to manufacture anything even though McCormick, Funk, and Legge had already scheduled production. Johnston's staff of 10 resurrected scrapped tools from McCormick Works and reconditioned them in a large circus tent they erected near his office. Mobile cranes moved massive tractor castings, some pieces so large they upset the cranes struggling in and out of the temporary plant in front of the McCormick Works. Winter arrived as the staff riveted tractor frames in the unheated tent and assembled wheels outside in the snow.

Johnston's staff completed 50 big twins before moving into their new home in January 1911, just after a New Year's Day winter storm destroyed the tent. The first buyers liked the powerful tractors officially named the Mogul 45s. The Mogul name came from sixteenth century Mongolian conquerors, but in England Mogul had appeared on several heavy-hauling steam locomotives. Reports soon came back that parts of the huge machine needed reinforcing due to the high-stress loads of large plow gangs and the intense, pounding vibration of Johnston's opposed two-cylinder engine. In an era when "engineers" without formal schooling combined the skills of blacksmith, inventor, and mechanic, Ed Johnston had become an accomplished problem solver. But what he did was not engineering, and some troubles were never quite resolved, even with his best educated guesses. Always in demand to solve problems not his own, always stretched thin with other obligations, Johnston was not having a good decade.

Gasoline prices, still 18 cents a gallon in the United States (about $2.74, adjusted to 2005 inflation), but almost double that in Western Canada, remained a concern. After initially rejecting the idea of distributing Rumely's Oil-Pull tractors, IHC began marketing them, especially overseas where farmers demanded kerosene tractors and where Rumely had no network. IHC expanded European sales. By 1910, to avoid protective tariffs on agricultural implements and tractors it sold in Europe, IHC opened factories in Sweden, France, Germany, and Russia. IHC was the first U.S. maker to sell a tractor in Russia, in July 1910. It marketed tractors in Argentina, South Africa, Austria, Mexico, Rumania, Brazil, Turkey, Italy, Uruguay, Spain, Peru, Switzerland, Chile, Norway, and Serbia.

IHC had competition now: 56 companies produced or announced plans to manufacture tractors in the United States in 1911, 14 more than 1909. Just after Thanksgiving in 1910, Legge asked his EC managers to look far ahead. Domestic sales managers pushed lighter tractors, from 8 to 12 horsepower. IHC's foreign sales manager advocated two forward speeds universally because in Argentina and Australia, where Steward's motor push binder and headers were already popular, Morton's friction machines were too slow. Yet everywhere John Steward looked in the room, he saw caution.

The engine for Johnston's Mogul Junior was one cylinder of the two he used in his Mogul 45. Bore and stroke remained the same, at 9.5 inches by 12.

"Looking a ways into the future," Legge said, "the light tractor business is the biggest thing before this company and ought to add at least 50 percent to the volume of business we are now doing." This motivated IHC's purchasing manager, Henry Utley, who added, "This development is so important that [we] could well afford to spend $50,000 to $100,000 working on a light tractor next season" ($750,000 to $1.5 million in 2005 dollars).

On May 1, a Special Tractor Committee, appointed by general manager Clarence Funk, officially met for the first time. It heard good news almost immediately. On May 19, the 25-horsepower Johnston single-cylinder tractor designed at Tractor Works appeared, weighing 2,000 pounds less than IHC's 25-horsepower Mogul tractor. It had a drawbar pull of about 4,500 pounds, and Johnston's engine ran on kerosene as fuel. The Mogul Jr., though not quite Steward's "light" tractor, was born.

At Winnipeg for 1911, IHC entered one Mogul Jr. in the "Light Weight" A class. This sole entry won the class by default. The trend still prevailed among all makers in 1911 to produce larger tractors. The winner there was Johnston's opposed twin, finally tamed. IHC's Mogul 45 produced 64.52 horsepower, pulled 6,650 pounds from the drawbar, and plowed 2.74 acres in an hour. Despite visionaries such as John Steward and Bert Benjamin, tractor manufacturers still meant for big tractors to make bigger power.

Titan and Mogul lines existing in nearly parallel development flew in the face of IHC's consolidation premise. The Deering-versus-McCormick rivalries died hard. The Sales Department still resolutely supported the bewildering assortment of similarly named, identically performing machines. Tractor Works churned out Moguls for McCormick dealers and Milwaukee produced Titans for the Deering branches, as a general rule; in communities with only one of the two IHC dealers, it might have carried Titans, Moguls, Morton friction drives, and IHC Gear Drives, or parts of all four lines. Events ahead would clarify confusion and end duplication within the next few years. The roles of Milwaukee Works and Tractor Works became clearer, and each week the engineers moved IHC's big tractors closer to perfection. For voices calling for smaller, more versatile machines, IHC seemed to move the wrong way. So the voices spoke louder.

Even though IHC's Moguls and Titans steadily improved, farmers began looking elsewhere. Canada's influence on tractor design had waned as attendance at Winnipeg fell. Manufacturers

1913 Mogul Junior. By June 1, 1911, IHC knew it had another success coming. Ed Johnston's latest idea, this 25-horsepower single-cylinder tractor, had come in weighing 15,400 pounds, about 2,000 less than its target.

produced almost 11,500 tractors in 1912; IHC built nearly 3,000 of them. Now, farmers who believed they needed a tractor wanted one to fit within their fences. Many Midwest and southern farms were only five acres and few were larger than 40. Farmers knew they purchased cast iron by the pound. Most recognized the value in working faster than a horse's walk, but they had already paid for their draft animals and those creatures ate off the land they worked. Small companies responded rapidly to farmers' desires and quickly slapped together workable devices from pieces gathered from various sources. To some makers, how well that tractor worked was inconsequential. When Clarence Funk's Special Tractor Committee met on January 6, 1911, its agenda was brief but not myopic.

While many of IHC's regional Sales Departments remained content with 20-, 25-, and 45-horsepower machines, the area east of the Mississippi hungered for smaller, lighter tractors. John Steward's ally Henry Utley suggested IHC buy one model of the Universal Tractor built by Northwest Thresher Company in Stillwater, Minnesota. While Titan and Mogul 45s weighed 20,000 pounds, the 18-horsepower, two-cylinder Universal tipped the scales at 9,000 pounds and featured sturdy 10-inch-long pistons mounted to a beefy 3.5-inch-diameter crankshaft.

Introduced in 1911, the Junior remained in production into 1913. IHC manufactured 812 of them.

However, William Cavanaugh, Ed Johnston's assistant, felt IHC should use the Universal as its next heavy tractor. Cavanaugh pushed for an even lighter model, of perhaps 5,000 pounds, offering four-wheel drive.

Johnston liked the four-wheel-drive idea. A week later he introduced sketches of a 6,000- to 8,000-pound machine using his 35-horsepower two-cylinder engine. Even Harry Waterman pronounced it "a good foundation and [I have] no doubt as to demand. But [pay] attention to the fact that in designing a general purpose tractor, we could not expect the same efficiency as obtained from [our] regular machines." Initially opposed to multi-cylinder engines, John Steward reversed his opinion. Now he advocated that IHC "build two sizes of tractors and use a four-cylinder engine on the larger size, cutting that in two and using half of it as a two-cylinder engine on the smaller machine. On this construction we would have to drive with all four wheels, or with that caterpillar plan."

Steward, one of IHC's most avid students of patents, saw reports of Holt Manufacturing Company in Stockton, California, and Northern Holt Company in Minneapolis. While the idea of crawlers had been present in farming for more than 20 years, the name "caterpillar" came from Ben Holt's machines. (Within five or six years of this meeting, 11 manufacturers produced crawlers or half-tracks. By 1922, that number doubled.)

Throughout the summer of 1911, projects came and went. Several European governments threatened to ban IHC tractors because of their noise; Deering Works developed mufflers for Titan tractors sold abroad under the Deering label. NWC chairman B. A. Kennedy asked both Waterman and Johnston to experiment with more efficient engine cooling. Ed Johnston, still competing with Waterman, reported having tested closed systems. "We must have an enclosed cooling system," he urged. "Two types of radiators are now coming through the shop and one is being made by an outside firm." Ten years after consolidation, engineering progress still came down to McCormick versus Deering.

More promising were April 15 discussions on a "general purpose tractor capable of pulling three plows, with not less than two or three speeds, maximum of seven miles per hour, weight between 5,000 and 8,000 pounds and retail for about $1,000" (roughly $15,000 in 2005 dollars). Discussion around the table suggested Engineering could produce something quickly by adapting

At the Winnipeg trials in 1911, IHC entered a prototype in the "Light Weight" class. As the sole entrant, it won by default.

the two-cylinder engine designed for the Cavanaugh-Johnston Tractor Works four-wheel drive on which prototype assembly had already begun. By mid-May, John Steward's "light general purpose tractor" had completed field tests. It resembled the Harpstrite traction engine and the Hackney Auto-Plow (which Tractor Works also had purchased for examination). Plows hung below the middle of the frame. Steward told the committee that he "viewed the farm requirements in a light tractor as precisely what is required of horses." His goal was to build a tractor that didn't waste so much of its engine power merely moving itself.

While Waterman and Johnston continued prototyping machines to compete against one another, a new candidate entered the contest. Deering Works hired Harry C. Waite, an independent designer. Waite devised a finely engineered lightweight machine that immediately went into farm testing near Lewiston, Montana. He deleted wheel and chain roller bearings, revised his three-speed transmission to two forward gears, and reduced the tractor's weight, cutting initial $1,500 manufacturing costs to $850 in quantity (about $22,500 and $12,750, respectively, in 2005 dollars).

John Steward adopted Harry Waite. Embracing the new ideas resulted in the Steward-Waite tractor, "as a three-wheel machine, two-speeds, 3.5 and 2.5 mile-per-hour, single lever control, to protect the carburetor air intake from dust by straining the air through water. All moving parts [are] to be absolutely covered. Also, to provide an attachment at a cost of $5 or $6 [roughly $75 to $90] so that any standard grain binder could be operated by pushing it before the tractor." Like all of the engineers, Steward promoted his pet ideas.

Second-Generation Power Farming Looks to Smaller Machines

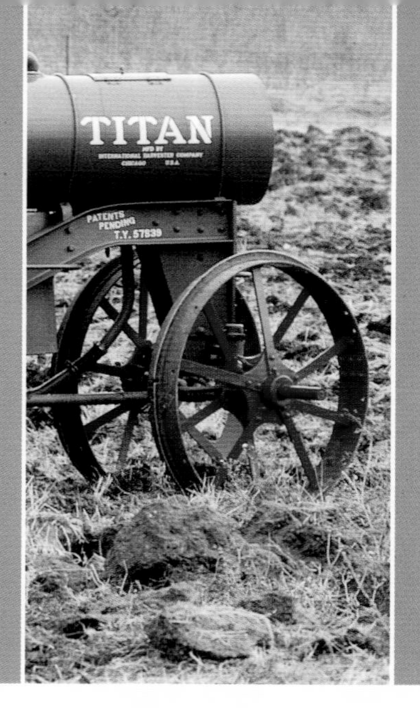

The International's short wheelbase captured the imagination of industries needing a machine to push or pull loads around factories and storage yards. Its early use as an industrial tractor helped sales.

Chapter 2

1914–1928

By the end of 1913, while there were 80 companies in the business, barely 50 actually produced just 7,000 machines, about a third less output than 1912. Sales problems and matters of public acceptance of the tractors were different issues. It was a demanding period for farm equipment makers and for farmers themselves. Following nearly three decades of hard times, from 1870 into the 1890s, farmers had come back. By 1879, farmers had recovered from the depression of 1873. Grain production increased as farmers fed a growing population, including a quarter-million emigrants fleeing famine and hardship in Europe and Asia. Abraham Lincoln's 1862 creation of the U.S. Department of Agriculture (USDA) acknowledged the significance of farming. The Grange and various Farmers Alliance movements from the 1870s gave voice to their concerns. These organizations briefly united farmers in demanding more equitable economic systems and regulation of the monopolies they thought took advantage of them. The Sherman Anti-Trust Act of 1890 was a direct result.

INTERNATIONAL
TRACTOR WORKS
CHICAGO
ILL.
HARVESTER CORP.

As the remaining unbroken prairies moved south, so did the significant tractor exhibitions. By 1914, the place to demonstrate new machinery had shifted from Winnipeg, Manitoba, to Freemont, Nebraska. Here a Titan 12-25 gathers a crowd.

Railroads reached the West Coast in 1893 and crops moved faster, farther, and cheaper than before. This opened trade around the Pacific Rim, bringing cotton and grain to Asia and returning profits to America's farmers. Prices rose steadily beginning in 1897 and they continued, sometimes taking dramatic jumps. In 1910, farm journals tentatively labeled the first decade "The Good Years." Yet, even as they saw "The Great Years" ahead, editors and writers raised a chorus of discontent. They urged tractor manufacturers to offer machines farmers needed: better built, more maneuverable, more reliable, easier to start, less cumbersome to operate, less costly to purchase. In their editorials, *The Country Gentleman* and *The American Thresherman* called out to IHC (which held one-third of the market by 1911), and to Avery, Rumely, J.I. Case, Minneapolis Threshing Machine Company, Emerson-Brantingham, and Hart Parr. Not only did these makers react, but dozens of other independents jumped into the business. By the end of 1914, there were 61 firms in the United States and Canada producing farm tractors. The sensation of a Golden Age was nearly universal. It was only a sensation.

IHC began eliminating products. Still, it was not major manufacturers who answered magazine cries for compact, lightweight tractors. Small makers like Bull Tractor of Minneapolis produced a 5,000-pound 5- to 12-horsepower tricycle. The Little Bull retailed for $335 (about $5,025 in 2005 dollars) and, by the end of 1914, the company sold 3,800 of them; IHC slipped to second place in sales behind the upstart. Robert Hendrickson and Clarence Eason, designers of The Wallis "Cub," advanced the benchmark and took credit for introducing the "unit-frame" tractor. In response, Alex Legge encouraged Addis E. McKinstry, the new manager of Experiments, to design and build a 5,000-pound, four-wheel, 8–16-horsepower tractor to run with a two-cylinder engine cast "en bloc" with mechanical valves. Legge hoped the two-cylinder engine would distinguish this machine from Tractor Works' Mogul 8-16, and intended "the work be rushed, as it is necessary to have a tractor of this size to satisfy local agents" who had to compete with Bull and Wallis.

IHC's implement genius Bert Benjamin spent some time on loan to Henry Ford to help the automaker develop implements for his Fordson tractor. While at Ford, Benjamin learned about assembly line production and "high-speed" automobile-type engines. Both of these factors came into play when IHC introduced its remarkable International 8-16. This is the engine assembly line.

Legge, McCormick, Deering, and IHC's Executive Council faced a dilemma. Dealers wanted product. Yet they still competed with each other more than with outside rivals. By mid-1914, the U.S. Justice Department had begun to pound the conglomerate. IHC had invested $28 million in new plants and new products ($423 million in 2005, adjusted for inflation) during its first eight years of consolidation. Sales rose from $56 million in 1905 to $101 million in 1910 (roughly $1.5 billion in 2005). Its assets, worth nearly $173 million (almost $2.6 billion in 2005), ranked it as America's fourth-largest company. George Perkins was a generous donor and influential Republican Party fund-raiser—so President Theodore Roosevelt blinked whenever he glanced toward IHC's size and tactics.

Roosevelt's successor, Democrat William Howard Taft, didn't inherit Teddy's IHC blindness. In 1912, the Justice Department filed suit against the company as its 10-year voting trust expired. The government charged that IHC "monopolized the harvester and binder markets, destroyed competition by forcing dealers to sign exclusive contracts, created a patent monopoly and reaped excessive profits."

IHC's gear-drive 10-20 and 15-30 models represented a big advance in tractor engineering, manufacturing, and machine reliability. Here a prototype photographed on July 11, 1925, shows off an experimental articulated rear-wheel cleat for traction in soft soils and sand.

IHC bore additional stigma. In 1895 Cyrus's brother Harold had married into the Rockefeller family. John D.'s Standard Oil Company fell under constant Justice Department scrutiny. It was too much for Taft to ignore. In 1913, the McCormicks, following advice from outside professional management, urged Clarence Funk (whom they always felt was a Deering loyalist) to resign. After Funk left to become president of Rumely, they promoted assistant general manager Alex Legge to the job. Legge first worked for McCormick in 1891 collecting bad debts for the Omaha, Nebraska, branch. He came equipped with good judgment, excellent business sense, and long-held loyalty.

Courtroom procedures occupied corporate time and resources; however, business, especially tractor business, went on. Farming tumbled from its profitability peak in 1910, sinking into a

Rural electrification as a national policy was still six years off when IHC showed this 1930 prototype. This gear-drive 10-20 carried an electric power take-off on the left rear fender beside the steering wheel.

recession that began with Wall Street banking worries in 1907 and bottomed out with a disastrous partial crop failure in 1914 (one that wiped out Funk's Rumely). IHC delivered 3,831 tractors in 1912, just 1,930 in 1913, and barely over half that, 1,095, in 1914. The corporation, pushed by its Sales Departments, still offered 16 different models in 1913; six of those sold fewer than 40 tractors each. In 1914, there were 17 models, but, ironically, only the giant Titan 45-horsepower Type D and the smaller Mogul 12-25 sold more than 100 each.

While appealing the anti-trust verdict, which bought them time, IHC's divisions prepared for dissolution forced by the courts. By early September 1915, an era of fierce internal competition passed. In late June, John Steward died of heart failure. He was 74 and had worked at Deering and IHC until two weeks before his death. Harry Waterman, Milwaukee Works superintendent,

1915 Mogul 8-16. IHC introduced this model in 1914. Its arched front framework allowed it for its time to turn extremely tightly within a 20-foot radius.

retired, and Paul Schryer replaced him. In Chicago, the Executive Council named Ed Johnston manager of the Experimental Department of Gas Power Engineering, in charge of design of all tractors, engines, and trucks. H. B. Morrow replaced Johnston as superintendent of Tractor Works.

Production considerations brought more changes on August 1, 1917, soon after the EC moved the Milwaukee Works Experimental Department to Chicago and consolidated it with Tractor Works Experimental. This combined group adopted the name Gas Power Engineering Department (GPED). It relocated some manufacture from Milwaukee Works to Chicago beginning with the 1918 model year, "to relieve the crowded condition of Milwaukee Works," caused by scheduling production of its new smallest tractor, the Titan 10-20.

On April 6, 1917, the United States entered the World War that had begun nearly three years earlier in Bosnia. By 1917, the Justice Department's anti-trust battle with IHC was nearly finished. Previously, the Justice Department had encouraged other anti-trust defendants to propose resolutions to their own suit, even letting them schedule actions to their benefit. IHC already had lost plants, sales branches, and products in Germany, France, and Russia either to nationalization or bombing before the United States joined the conflict. McCormick offered to sell its three old-line harvesting machinery subsidiaries. (This proved a benefit because harvesters were its corporate loss leaders.) Cyrus Jr. suggested reuniting the divided companies as International Harvester Corporation once again. On October 12, 1917, Ed Johnston noted that IHC had changed the name of the small Mogul tractor to 8-16 International in anticipation of a federal ruling.

The horizontal single cylinder ran a planetary gear transmission to power the rear wheels using a left-side chain. The transmission provided only one forward and one reverse speed.

1917 Mogul 10-20. As successful as the 8-16 was (and IHC sold something like 5,000 of them in 1915 and more than 8,000 in 1916,) Ed Johnston felt they needed more. He increased engine-operating speed and added a second forward speed. IHC offered optional plow guides for 8-16s and continued with the new 10-20s.

Still narrow, at 56-inches, and short at just 5 feet, the twin-cylinder 10-20s appeared mid-1916, overlapping with the popular 8-16. Sales of the 10-20 never matched the earlier model.

Prior to peace in Europe on November 11, 1918, IHC surrendered to the U.S. Justice Department. The most painful terms of its agreement were stipulations to own and operate only one sales facility in each town. IHC released 4,778 dealers. Most of them quickly joined John Deere. That was the bad part of IHC's surrender. The good part was very good: IHC could make use of all those pooled funds and facilities to strengthen its growing tractor business.

THE RIVALRY THAT ALEX LEGGE CREATED between Ed Johnston's Chicago Tractor Works and Harry Waterman's Milwaukee Works was a highly advanced, forward-thinking strategy. Two separate design teams developing similar products is a technique common at the beginning of the 21st century. Milwaukee Works began designing the two-cylinder Titan 10-20 in late 1914 as one of Waterman's final challenges to his younger rival, giving the Titan slightly more power than Johnston's 8-16 first Mogul (one cylinder.) The beginning of war in Europe increased demand for food and textiles, a need met in the United States by farmers attracted to smaller machines. In 1915, IHC sold 5,841 tractors; more than 5,000 of those were Johnston's 8-16. (Bull Tractor had not stalled; it introduced the 7- to 20-horsepower-rated Big Bull, weighing 4,500 pounds and selling for $585, roughly $8,800 in 2005 dollars.) For 1916, Ed Johnston, now manager of experiments for the GPED, increased cylinder bore and replaced the single-speed forward transmission with a two-speed unit, renaming it the Mogul 10-20. The EC approved production in November, one month after blessing the other 8-16 four-cylinder International Mogul on October 16. IHC wanted this Mogul to share an engine with the Model G truck, but oil leakage problems delayed introduction both for the truck and tractor, which, as IH historian Guy Fay reported, Engineering set aside until it had completed design and fully tested its own engine.

1918 International 8-16. This was a great idea that led to other great ideas, including power take-off and chain final-drive systems. But its own channel frame was not strong enough and this tractor never was a great success.

This other 8-16 tractor represented a tremendous advance for IHC. Its compact dimensions made it maneuverable and practical for small-acreage farmers. Its tapered nose appealed to orchard operators and its shape, reminiscent of contemporary automobiles, brought 8-16s to the attention of industrial users. It was too radical for some conservatives, but for those buyers, IHC had more traditional 10-20 Mogul and Titan tractors. When Henry Ford rose on the horizon as the next major challenge to IHC's peace of mind, it was the unremarkable Titan that fought off Dearborn's assault.

On October 7, 1913, Henry Ford's first Model Ts drove off the assembly line at his Highland Park, Michigan, plant. His workers produced a car in three hours moving along the 250-foot path. Significant as this was, this was only one of Ford's warning shots fired toward other U.S. manufacturers. For tractor makers, the first signs of war appeared as early as 1905 when Ford experimented with prototypes called "Automobile Plows" on his farm in Dearborn. His chief tractor engineer, Joe Galamb, worked for another decade on numerous designs and variations.

When Galamb learned about Hendrickson and Eason's unit-frame Wallis Cub tractor almost a year before its introduction in 1913, he and Farkas went into a productive frenzy, creating what became the Fordson in 1916. With this small, lightweight, unit-frame tractor as his ammunition, Henry Ford declared war on the horse. Every other tractor maker wanted to defeat draft animals as well, and they all became victims of the assembly-line juggernaut called Henry Ford & Son.

IHC's four-cylinder engine ran at 1,000 rpm using a magneto for ignition spark. Early engines suffered from inadequate lubrication.

It was a clever concept with the radiator and fan behind the engine and below the fuel tank. Engine heat helped vaporize the fuel. It probably helped vaporize operators' feet as well.

Even though Ford claimed he didn't care about other tractor manufacturers, he competed against International Harvester and everyone else for the farmer's dollar. In 1916, there were 165 companies claiming to be tractor makers. Some of these continued producing simple variations on their earlier heavy, ponderous units, with an occasional All-Purpose or Motor-Cultivator–type machine introduced to respond to changing times. Others, inspired by Ford's well-publicized efforts to transform his Model T into a tractor, tried similar, automobile-based attempts. A core group, including IHC, followed a middle ground, taking from both extremes the technologies of value and ideas with merit.

As world war became inevitable, overseas needs for food and cloth increased farmers' profits but also induced them to increase production. The war drew nearly every available horse and

1919 International 8-16. The tapered nose appealed to orchard owners who saw its ability to slip under branches. Its radical looks however, put off some conservative buyers at the same time it reminded others of contemporary automobiles.

The 8-16s were IHC's first products manufactured on assembly lines. Both engines and tractors advanced along a moving conveyor during assembly.

able-bodied male to Europe. On battlefields in France and Germany, a horse's life expectancy was 7 to 10 days. In 1900, there were more than 23 million horses on farms in the United States. In 1910, there were a million more horses, yet there were only 3,650 tractors at work. The value of tractors became apparent to those at home. Farm journals pushed for useful small tractors and city newspapers picked up the tractor idea, engaged by its novelty.

In 1850, 70 percent of the U.S. population worked in agriculture; it was 33 percent by 1910. The population had increased, yet fewer people fed and clothed the others. In 1917, England planned to convert all green lawn to food production. There were just 600 tractors in the United Kingdom then. By the end of 1918, there were 3,000, and a year later nearly 7,000 in all had opened up almost 1.4 million new acres to cultivation. In France, 4,000 mostly American-made tractors worked to restore farming land damaged by war. To plow an acre of land behind a horse meant the farmer walked nearly seven miles, taking as many hours. The USDA reported 34,371 tractors working on U.S. farms in 1916. In North America, 70 active companies (of the 165 claiming to be in business) manufactured 29,760 tractors. Nearly 80 percent of these were two-plow machines rated to work at more than 2 miles per hour. A tractor towing 2, 3, or 4 plows proved amazingly more efficient to newspaper writers who had never seen steam traction engines pull 15 plows. Horses that sold for $49 in 1900 ($735 in 2005 dollars) brought $109 in 1914 ($1,635 in 2005 dollars) and $139 in 1918 (nearly $2,100 in 2005 dollars). If bombs and rifle shots didn't get them in France, their hearts failed under the load of heavy plowing in Kansas or Georgia or Maine. Tractors would save America, the editors wrote, and they could save the world.

The 4-inch-bore and 5-inch-stroke engine went through several incarnations as IHC struggled to get its problems resolved. The tractor ended up 600 pounds heavier than the similarly powered Fordson, and it cost more, further harming sales.

North American makers turned out 62,742 machines in 1917, shipping 14,854 to Europe. In 1918, wartime uses of steel and other material threatened all domestic industry. The U.S. Government Priorities Board began to speak of rationing steel, limiting first-generation development prototypes to 10 units, second-generation test models to 50, and total production by all manufacturers to 315,000 tractors, hoping this would be enough to continue to feed and clothe the world. In fact, the final count reached only 132,697, and the armistice in November brought an end to allocations in December. More than 100 new companies, seeking to ensure the right to profit from the war in Europe (or to continue receiving adequate supplies of other raw materials), entered the tractor business, bringing the census of supposed makers to more than 250. Of these, 95 responded in a USDA questionnaire to essential industries that they had produced even a single tractor in 1918.

SITTING ON IHC'S NEW WORK COMMITTEE or Executive Council in those days was akin to living in nearly constant turmoil interrupted by brief calm. The corporation's bold new channel-frame tractor, the four-cylinder Mogul (later International) 8-16, conceived in 1914 and first born in August 1916, was reborn later in 1917. Its problems required two engine transplants. The first enlarged the crankshaft diameter to strengthen the engine and also remedied failures

1920 Titan 10-20. It lacked the sophisticated appearance of International's 8-16, but the Titan became the standard bearer for IHC when Henry Ford and his Fordson declared a price war. With new machines in the pipeline, IHC's Alex Legge could afford to discount Titans and match Ford price, cut for cut.

involving splash lubrication. The second addressed horsepower inadequacies. It always used in-line vertical four-cylinder engines, coupled to three-speed gear transmissions driving the rear axle by chains. As the International 8-16 (the name ultimately given it on October 12, 1917), it offered farmers America's first production power take-off (initially tested on 50 prototypes and approved April 4, 1919). It provided the GPED with a test vehicle on which to experiment further with four-wheel and six-wheel drive. Engineers developed crawler tracks, and in June 1919, extra width, reinforced steel wheels appeared for rice fieldwork. In addition, the International 8-16 developed and proved IHC's own assembly-line production beginning in 1918.

By then its cousin, the Mogul 10-20, was approaching retirement and the Titan 10-20 should have been. IHC discontinued the Mogul in 1919, yet it continued the Titan until 1922 for a number of reasons. Ford's competition, and frustrations with the smaller tractor, left the company little choice. The 8-16's lubrication system led a small tube from the transmission case to the external final-drive chains. It vented to outside air, allowing dense, fresh air to mix with warm lubricating oil vapors inside the case, which sometimes ignited, fracturing the case. Legge considered substituting a gear final drive already in development, approving this version on September 27, 1920. But even this decision was only a Band-Aid to a severely injured project.

continued on page 48

This was another successful machine for IHC. In 1920 alone, the company manufactured 21,503 and output from 1915 through 1922 totaled 78,363 units. This was a 3-plow-rated machine compared to the two-plow-rated Fordson.

The Titan's two 6.5 by 8-inch cylinders gave the tractor enough power for three plows. Its two-speed gearbox provided a choice of 2.25 or 2.875 miles per hour forward, and a single 2.875 miles per hour reverse.

1921 15-30 Gear Drive. IHC was not the first tractor maker to utilize unit-frame technology. But the single large casting guaranteed extreme rigidity and complete enclosure that enhanced engine and gearbox life and offered much greater tractor strength.

It was a simple yet capable tractor. The boxed-housing through the middle of the operator's platform contained IHC's power-take-off shaft. This was a remarkable feature for 1921.

When IHC's executive committee first approved these for production in late 1920, the Naming Committee saddled them with an impossible name: "12-25 Four Cylinder International Tractor with gear-drive, burning kerosene."

Continued from page 42

Johnston and Morrow warned that 8-16 channel frames were too flexible to hold tolerances necessary for severe loads. Under peak stress twisting, frames allowed gears to disengage. Johnston ordered a complete redesign on January 20, 1922, calling for a "Platform made in one piece [a unit frame] and heavier metal," and a half-dozen other supports, reinforcements, and enlargements meant to strengthen and stiffen the tractor. Then, instead of spending the money to make new casting patterns and to revise the production line that assembled the 8-16s, those efforts and resources went into two new machines in a new line.

The GPED approved production of the first of these new machines, initially authorized as the "12-25 Four Cylinder International Tractor with gear final drive, burning kerosene" on October 1, 1920, shortly after the Gear Drive decision for the 8-16 (eventually becoming the McCormick-Deering 10-20). Almost on the eve of production, the Naming Committee changed its designation to the International 15-30 (later renamed again as the McCormick-Deering 15-30). In its second year of life, Milwaukee introduced a vineyard version. At the request of the Los Angeles and San Francisco branch houses, a "Special California Type" came immediately afterward, fitted with 12x40-inch rear wheels and 24-inch-diameter fronts without the belt pulley or radiator curtain "unless specially ordered as an attachment."

When the original International Gear Drive 15-30 arrived in 1921, it adopted several pages from Henry Ford's lesson book. First, the new tractors were, like the Fordson and the Wallis Cub, unit frame modular construction machines. The International, renamed the McCormick-Deering 15-30 on August 22, 1922, and the McCormick-Deering 10-20 introduced in 1923, housed the engine, transmission, and gear final drive, all preassembled elsewhere, in the single-piece cast-iron tub onto which production line workers bolted front axles and rear half-shafts. (The Executive Council retained the International name "for use on tractors supplied to Foreign Countries where the name International is preferred," and possibly in response to the government's reinstituting anti-trust proceedings that July.) The following September 4, 1923, Ed Johnston signed off on the "increased power" 15-30, rating output at 22–36 horsepower. The tractor designation remained 15-30 despite Sales Department efforts in mid-1930 to order "name plates, transfers, stencils, and literature as for the model 22-36." However, their undated handwritten change order went unsigned, and the only 22-36s labeled as such were those meant for export.

A similar pattern followed the smaller 10-20, which IHC first authorized for production on August 25, 1921. As with the 15-30, it carried a power take-off from the transmission as standard equipment. On September 2, 1922, just in time for the first production run, the Naming Committee retitled the machine a McCormick-Deering for sales in the United States and Canada, while keeping the International trade name for foreign sales. This relatively orderly progression of prototype development, testing, eventual mass production, continued evolution, and improvement remained measured and deliberate. However, disorder and improvisation followed a parallel path behind the scenes.

Experimentation continued with few breaks. While Ed Johnston had seen to it that the PTO appeared on both 15-30 and 10-20 models, he had barely improved its performance or capabilities

The adjustable rear-hitch device was an option. IHC advertised these as triple-power tractors, with drawbar, belt-pulley, and PTO as power-output options.

IHC used Waukesha to assemble engines in this and other early 15-30s. Then for 1923 production, the corporation began manufacturing its own engines with 4.5 inches of bore and 6 inches of stroke.

In 1928, Ed Johnston authorized the Increased Power version of the 15-30. With bore enlarged to 4.75 inches, power output went up to 22-36. Tractors retained the 15-30 designations in the United States, while overseas some appeared labeled as 22-36s.

from his earliest efforts. There had been little economic impetus to push development. That effort came from Bert Benjamin in 1917.

Benjamin had graduated from Ames College in Iowa with a degree in mechanical engineering and went straight to work for McCormick Harvesting Machine in 1893 as a 23-year-old draftsman, working in the McCormick Works Experimental Department where he remained until 1899. He spent that year in manufacturing as chief inspector before going back to the Experimental Department until early 1903, when he returned to manufacturing, again as chief inspector. Then, in 1910, Alex Legge named him superintendent of McCormick Works Experimental Department. Throughout the years, Benjamin specialized in implement design.

In September 1917, Benjamin was in Napanee, Indiana, watching three of his International 4-horsepower "Binder Engines" harvest hemp, a product in great demand by the military for a variety of uses. A Titan 10-20 pulled one binder, four horses pulled another through medium-tall plants, and an 8-horsepower competitor pulled a third through very tall hemp. Benjamin watched hemp dust and leaves choke the auxiliary engine powering the towed binders. They lost perhaps 25 percent of their power. In soft ground, the nose of the binders sank under the weight of its engine. Tractors that ran ahead of the binders suffered no ill effects from the dust. The time wasted cleaning the binder's radiators and air intakes and the duplication of fuel costs led Benjamin to some conclusions about how harvesting might improve if the free-breathing tractor engine also drove the binder by power take-off shaft. Nearly a year later, October 21, 1918, Benjamin's McCormick Works tested a prototype PTO, operating the cutter bar of a mower attachment and a sweep rake lift. Its performance encouraged the orderly progression of development work that IHC had done with so many other products.

AFTER HENRY FORD HAD SUPPLIED ABOUT 4,260 Fordsons to England's Ministry of Munitions, he began producing them for American farmers around April 23, 1918. He still owed the English and Canadian governments another 2,740 tractors, but he felt a strong desire for the same recognition at home that he had achieved abroad. To that end, while the first 3,200 tractors left Dearborn, Michigan, with no identification on them, Ford & Sons stenciled its name on those that followed.

Ford's self-promotion galled his competitors, who protested to the Ministry of Munitions and the U.S. government against Ford's proposal to be the official tractor of the war, to provide one standardized machine to the exclusion of all other manufacturers. They accused Ford of maneuvering public sentiments to ensure his own uninterrupted supply of ore and coke to make his steel.

All the publicity created a demand for the small Fords, a hunger that Ford planned to satisfy through State War Board allocations. But Ford's own U.S. production got off to a slow start and other manufacturers began to fill farmers' needs. Congress passed the Food and Fuel Control Act to regulate prices and guarantee minimums for all commodities. The winter of 1918–1919 was mild, enabling farmers to work later in the fields in the fall and go in earlier the next spring. Those farmers with horses felt no need for tractors because of the luxury of good weather and

McCORMICK-DEERING

Both 15-30 and 10-20 engines offered removable cylinder liners and pistons spun on twin main-bearing crankshafts. The engine measured 4.5-inch bore and 5-inch stroke.

plenty of time. Those without enough horses soon were surprised when the Department of War sold off thousands of tractors after the armistice. Europe still needed food and cloth, and farmers, thinking the wartime economy would never end, expanded throughout 1919 and 1920.

During 1920, the USDA reported that U.S. manufacturers sold 162,988 tractors. It was the biggest year yet. Fordsons constituted 35 percent of these, while 15 percent came from IHC. Tractors worked on 229,334 farms that year, representing just 3.6 percent of all operations. To Ford and others, that meant the remaining 96.4 percent were potential customers. Then crop prices fell throughout 1921 as European farmers got back to work. Tractor production, except among a few firms, virtually stopped. By year-end, manufacture dropped to barely 35 percent of the peak reached during 1920. In January 1922, Ford had cut his Fordson's price by $165, from $790 to $625 (approximately a $2,315 reduction, from $11,060 to $8,760, adjusted to 2005 dollars). IHC followed during the spring and summer, dropping the $1,200 Titan 10-20 to $900 ($16,800 to $12,600); the $1,150 International 8-16 also went to $900—from $16,100 to $12,600—and the 15-30 decreased from $2,300 to $1,750—$32,200 to $24,500, in 2005 figures). General Motors entered the business through its Samson Tractor subsidiary, but by year-end Ford had sold half of the 34,000 tractors produced, IHC had sold one-quarter, and GM's sales were nearly negligible.

Horses still were cheap and crop prices remained low. During the winter, farmers repaired old equipment or sat on their hands. The market stagnated until February 5, 1922, when Ford, determined to keep his new Rouge tractor plant running, cut the Fordson price $230, from $625 to $395 ($3,220, from $8,750 to $5,530, adjusted to inflation for 2005). IHC followed, reducing the 8-16 another $230, and the Titan down to $700 ($9,800). To make his tractors look as good a bargain as Ford's, IHC president Alex Legge threw in either a plow or other implement. All the

Opposite
1926 10-20 Gear Drive. Following the 1921 International 15-30, in 1923, IHC changed the tractor names to McCormick-Deering and introduced the companion 10-20. This tractor also used unit-frame technology with geared final drive.

The 10-20 stood 14 inches shorter than its big brother 15-30. The smaller tractor also was more than 2,000 pounds lighter.

other manufacturers fell in line with industry-wide reductions averaging the same as Ford's $230 price cut.

These lower prices moved farmers off their hands. However, the late and very wet spring had an equal impact. Farmers who had hoped to get fields seeded using horses were delayed so badly that only tractors could get the fields worked in time to make a good harvest. In that same season, Ford made one of his few mistakes. Early in the year, he announced that his company would become a full-line producer offering a complete assortment of Fordson implements. Within a few months he reversed the decision, ordering dealers to select for themselves the implements they sold. It left Fordsons vulnerable to equipment mismatches that damaged their reputation. Yet it was GMC that quit first in 1922, having "invested" $33 million (nearly $500 million if adjusted to 2005 dollars) in a business it had entered specifically to challenge Ford.

In 1923, Ford reached its peak, producing 101,898 tractors and capturing 76 percent of the market. It left 9 percent, or 12,057 machines, to IHC. All the other 73 producers combined made up the remaining 15 percent. In 1924, Ford slipped slightly, to 70.9 percent, or 83,010 tractors, while IHC grew to 16 percent, with 18,758 machines. Competitors failed rapidly and dozens ceased production during 1924 because they either produced inferior machines or they made good equipment but could not afford the losses that Ford endured. Those who remained only garnered 14 percent of the total.

IHC dealers practiced aggressive salesmanship, seizing every Fordson demonstration as a chance to show how limited it was compared to the IHC tractors. This, and a still-suffering economy, further eroded Ford's market share in 1925. By year-end, he still had sold 64 percent of the annual production, but IHC had risen to one-fifth of the 164,097 produced. Ford chose not to improve or update his tractor, preferring only to get them into the hands of former horse farmers, "to ease the drudgery," he said. The Fordson revolutionized farming, proving that a small, lightweight tractor could be mass-produced, sold cheaply, and that it could replace the horse. IHC used the period of the World War and Ford's tractor war as a time of further experimentation. Only one of the efforts, a Motor Cultivator originating in 1915, did not succeed.

In 1927, the U.S. Supreme Court refused to reopen the Justice Department case against International Harvester, ending one of the two great challenges to the company's existence. Between New Year's Day and June 4, 1928, Henry Ford had assembled barely 8,000 Fordsons in Dearborn. On that June day, he ceased U.S. operations and transferred all Fordson tooling and manufacture to Cork, Ireland. His aggregate production during 1927 and 1928 gave him 31 percent of the market. In 1927, IHC claimed an equal amount, and in 1928 the company regained 62 percent of the business. That year, experiments begun long before started paying off. IHC introduced its TracTracTors, which were crawlers based on the McCormick-Deering 15-30 and 10-20. However, it was another experiment that dated back to the Mogul, that evolved from the unsuccessful Motor Cultivator, and that received an evocative and catchy product name, the Farmall, that IHC introduced in 1924. After a measured start-up, it sold 24,899 units in 1928. It revolutionized farming as thoroughly as the Fordson. IHC's Farmall then went ahead and plowed Ford under.

McCORMICK-DEERING

From Innovation and Disappointment to Innovation and Success

1929 Regular. Sporting the Texas sand wheels, this tractor is a long way from home now in its collection in Pennsylvania. Once IHC began offering Fairway models and Narrow Tread variations, it began to name the original Farmall "the Regular."

Chapter 3

1915–1932

John Steward and Harry Waite together had planted a seed inside International Harvester Corporation. The two men envisioned a machine capable of doing all the jobs routinely performed by horses, a concept encouraged in farm magazine editorials. Steward's successor, Ed Johnston, and his assistant, David Baker, developed the general-purpose Mogul. While other engineers redesigned horse-drawn grain drills, harvesters, and other implements to work behind the Mogul, it was still too clumsy and bulky to do delicate, precise row-crop cultivation. Baker helped Johnston, Carl Mott, Philo Danly, and John Anthony complete something they all called the Motor Cultivator in late 1915. Baker added a draw bar, believing that cultivating machines had possibilities beyond a single purpose.

Bert Benjamin at McCormick Works liked multiuse machines, yet he appreciated the Motor Cultivator's simplicity. Raised on a farm, he had done fieldwork from age eight until he left Iowa, engineering degree in hand, to join IHC. Benjamin's assignment at McCormick entailed adapting horse-drawn tools with wheels to power farming. His college education taught him to understand why wheels might not be necessary for tractor-drawn implements.

FARMALL
FARMALL

Tractor Works operators tested the Motor Cultivator through the summer of 1916 at an IHC farm in Aurora, Illinois, centering it in the rows with foot levers that shifted the spindly front wheels. To turn, they released a lock and cranked the steering wheel. This rotated the entire engine, transmission, and drive-wheel assembly.

The engine sat over the rear drive wheels behind the operator's head. Motor Cultivators overturned easily on side hills, so Baker and Mott added weights to the front wheels as counter balances that also made pedal steering harder and less responsive. This was "cut-and-try" engineering, seat-of-the-pants work of the type that Ed Johnston and David Baker knew best. Johnston's Motor Cultivator was the focus of many NWC meetings. The Sales Department liked the idea and, based on past successful Johnston designs, they argued that this one would work as well,

1920 Prototype Corn Harvester. Tractor Works engineer David Baker worked with three colleagues to create the Motor Cultivator in late 1915. While it never succeeded completely, it became the test bed from which important ideas grew. By December 1920, the machine had shed its cultivators and adopted a double-row corn harvester from Bert Benjamin's McCormick Works.

thereby avoiding normal routine testing. The NWC ordered 300 from Tractor Works for 1917 after testing only two or three prototypes. Baker, Danly, and Mott continued working on it as manufacture began. Tractor Works completed 58 by July, too late to be useful for the 1917 season. IHC halted production at 100 units; however, only 31 were shipped. On August 22, the NWC ordered 300 more for 1918. Then, in early September, Addis McKinstry, IHC's vice president for sales, recalled 1917 models to Tractor Works. Baker and the others changed the engine's governor and cooling fan, and they refitted heavier, cast-iron front wheels.

For the first time, IHC had used its customers for final testing. Johnston had a good reason for hurrying production. Neither stupid nor malicious, he anticipated shortages in raw materials because of World War I in Europe. U.S. involvement was inevitable by 1917; IHC reasoned that keeping existing machines in production was easier to justify than introducing a new one when rationing hit raw materials. Still, early buyers had found the machine was top-heavy, prone to overturning, its LeRoi engine was underpowered, and it was geared too slowly. This put Baker in a difficult position. A more powerful engine would add weight where they least wanted it, further jeopardizing its balance.

Henry Ford had influenced IHC's fate in 1917. As he readied Fordson's production for American farmers, he wanted implements. Ford considered his real competition to be draft horses, so he asked IHC (and Deere & Company) for advice. Alexander Legge, directing the U.S. War Production Board, seized the opportunity. He sent Bert Benjamin to Dearborn, loaning his Deering Works chief to Ford for several months. Benjamin studied Ford's assembly-line production and its high-speed automotive engines while he helped revise Fordsons to use McCormick-Deering equipment, and he designed a Fordson line of implements.

1921 Prototype International Steam Tractor. Fuel prices have concerned farm equipment manufacturers for a century. In the early 1920s, as gasoline prices soared above 20 cents per gallon, IHC considered reintroducing steam-powered tractors to cost-conscious farmers. This photograph, made on May 24, 1921, shows a prototype "light steam" tractor.

1924 Farmall Regular with Cultivators. IHC went through several experiments to locate cultivator mount points to be effective and easy to use amidst young plants. Introduced in 1924, IHC charged $825 for this tractor plus another $88.50 for the cultivators. Adjusted to today's inflation, that would be about $11,500 for the tractor and $1,195 for the cultivators.

Back in Chicago in late 1917, Benjamin invented a kit to transform Fordsons into cultivating tractors. Then Ford dropped plans for implements, leaving selection to his dealers. This meant frosting on the cake to Benjamin and Legge; now IHC would sell Ford dealers implements too. Motor Cultivator progress idled in place.

Sales liked the Motor Cultivator, but Manufacturing didn't. The economy tightened. Manufacturing's Harry Utley endorsed a two-tractor, large and small–machine plan. The Motor Cultivator was the distant third. Utley never produced enough tooling to meet a 300-machine order. The 1918 machines incorporated the changes retrofitted to 1917s, allowing higher engine speeds without overheating. The new model, with more parts made out of cast iron, weighed

1928 Farmall prototype with Plow Guide and Engine Cover. This Farmall shows off a few options, some destined for production, others not. IHC had marketed plow guides for its tractors for decades so the "Steering Device DL-2517" is no surprise. The single front wheel and the engine cover are unusual. This photo, dated March 12, 1929, shows one of IHC's 1928-year test "mules," #23397.

3,400 pounds (instead of 2,200) with a PTO, a belt pulley, and a hitch—all elements of IHC's new "Triple Power Plan." Engineers tested mowers and a sweep rake–lift mounted on units without cultivating blades. Tractor Works assembled 160 by July 15, ending the year at exactly 301, including the 1917 rebuilds. This slow pace resulted from steel shortages during World War I; the government allowed manufacturers 75 percent of 1917 quantities. It wouldn't review allocations until December 1918.

On August 22, about 10 weeks before the armistice, the NWC reduced production for 1919, knowing IHC would get less steel. McKinstry calculated it cost $500 to manufacture each Cultivator (approximately $7,500, adjusted to 2005 figures) plus transportation, commissions, sales costs, and overhead against a $450 retail price ($6,750 in 2005 dollars) for Avery's Cultivator. A week later, IHC terminated 1919 production. They delivered 213 in 1918 (67 leftover 1917s redone as 1918s) and 84 in 1919. The last 62 units sold for $450 each, a $50 loss ($750 in 2005 dollars) through mid-July 1920. The final report, written charitably toward Johnston and

1924 Farmall Fairway. Almost from the start, IHC marketed Fairway models of its Farmall Regular tractor. These came with especially wide steel wheels to float over soft lawns at golf courses, parks and large private estates.

Tractor Works, judged the Motor Cultivator "could not be produced at a cost which it was estimated the farmer would pay." In 1920, IHC wanted farmers to buy not only a "general purpose" tractor (large or small) for plowing and harvesting but a second machine as well, for cultivation.

On June 11 and 12, 1919, Johnston had taken Motor Cultivators to Blue Mound, Illinois, for the first strictly cultivator show in the Corn Belt. Avery, Moline, and Allis-Chalmers demonstrated machines. J.I. Case Plow Works introduced a rail-frame 12-horsepower cultivator with two tall drive wheels at the rear, steered by a single wheel in front, its engine and cultivators in the middle. Far lighter than a unit-frame Wallis Cub, it signaled a new direction in machinery design. About a month after the Illinois show, Bert Benjamin sent a memo to Alex Legge on July 27, 1919. Benjamin described "the next step . . . toward getting considerably more production from one operator."

Benjamin's memo envisioned a highly adaptable "Combined Tractor Truck" of two- to four-ton capacity, powered by a 15- to 25-horsepower kerosene engine. It would carry "a combined harvester-thresher, power direct from the engine, with speed independent of the tractor, could carry and operate a grain binder and shocker, a corn picking device with a box for loads, a hay loader and rack for hauling hay, or a water tank, pump and sprinkler device for fire protection on the farm." Benjamin described something fanciful. Alex Legge was more practical. He

The 200,000th 10-20 Gear Drive. On June 4, 1930, production stopped for a few moments for a celebration. Despite dealer fears that Farmall sales would sour the market for standard-front 10-20 and 15-30 tractors, IHC manufactured this landmark wide-front machine sixteen years after Farmall production began.

insisted that whatever this "next step" did, it "must at least look like a tractor." The conservative Scot understood that Midwest farmers might shy away from something too different.

Although IHC had stopped manufacturing the Motor Cultivator, Mott and Johnston continued development efforts. In early 1920, they reoriented the engine to the direction of travel and moved it slightly forward on the frame. This immediately improved the balance. They left the narrow-end rear wheel to do the steering, but using a differential with sprockets and chains, now the engine drove front tractor wheels. They introduced the automatic differential brake. Benjamin set cultivators closer to the rear single steerable wheel. They referred to this as the "Cultivating Tractor." In late 1919, Johnston, seeking to better differentiate this latest version, asked for name ideas. Ed Kimbark suggested "Farm-All," recorded in the

1918 Motor Cultivator. Tractor Works engineers tested these throughout the summer of 1916. They learned the machines overturned easily on side hills, causing David Baker and Carl Mott to add counterweights to the front wheels.

November 10 Tractor Works records. By early 1920, they dropped the hyphen and the name became Farmall.

Benjamin had nearly completed his long-term study of farming methods and implement design. He took his research and ideas to Tractor Works. Johnston assigned Danly, Mott, Baker, and Anthony to design tractors for Benjamin's new implements. One, designed with David Baker, used a Waukesha engine fitted with a reversible operator's seat and a transmission with three speeds, both forward and reverse. These new ideas got little support. The Depression, slow sales, and Motor Cultivator losses left little money to hand-assemble more than two Farmall prototypes and test them with their own implements. Prototypes are colossally expensive, often costing 50 times what a finished production version can cost to manufacture. The Model A Farmall, using an L-head truck engine from Akron, appeared around February 7, 1920; the Model B arrived June 30, using a new engine that carried through to production.

Benjamin campaigned for the reversible Farmall. He wrote Legge on October 15, 1920, that prototypes performed 11 separate farm operations using a single operator. He adopted automotive-type engines from 8-16s, pricing his machine for $900 (roughly $12,600 adjusted to 2005), while the International 8-16 at $1,000 (about $14,000 in 2005) did only four tasks with one individual who would have to rely on horse teams to do the other seven jobs. Johnston surprisingly announced support for the Farmall. "Farm power, increasing production per man," he said, "was the coming thing in agriculture industry." Harry Utley didn't believe all-purpose machines even were desired and thought farmers wouldn't sacrifice good equipment to switch.

Above left

The four-cylinder engine developed 12 horsepower at 1,000 rpm. Cylinder bore was 3.125 inches with 4.5-inch stroke. The entire motor cultivator weighed just 2,200 pounds.

Above right

Farm operators were kept busy centering the cultivators between rows using foot levers that shifted the front wheels. Conceived to cultivate corn, IHC hoped to compete with Avery and Moline who had fully developed models on the market.

Johnston warned that running a modest experimental program would cost $150,000–$300,000 to build five tractors and implement sets by hand. (This is equivalent to between $2 and $4 million in 2005.) Knowing that doubling the number of prototype tractors would not double the costs, he advocated enlarging their test fleet. J. F. Jones, the Chicago office sales manager who soon would become the Farmall's arch foe, felt that farmers wouldn't take to it because it was "built on exactly the wrong lines." He suggested replacing it with something heavier.

Legge remained neutral. He operated a farm about two miles north of the Hinsdale facility. He and fellow "hobby farmers" Harold and Cyrus McCormick recognized the Farmall's value. Still, cash was tight; IHC remained at war with Ford and after the Motor Cultivator disappointment, the only way the McCormicks could support this program was slowly and from deep in the shadows. Benjamin became an apostle. He wrote letters and he cornered anyone he encountered with any influence. He converted Johnston's staff to the machine, its implements, and the work it could accomplish.

Around Christmas, one of Benjamin's engineers, C. A. Hagadone at McCormick Works, sketched a lighter-weight version of the Farmall at about half the 4,000-pound approved model. On January 21, 1921, the NWC cancelled the five heavy prototypes and ordered "two of the modified, lightened Farmalls." This was the first time Ed Kimbark's Farmall name appeared in official IHC papers.

The lightweight Farmall ran in one direction only, steering in front while its two powered wheels pushed from the rear. By May 1921, Tractor Works had enclosed final-drive housings and

By September 1917, IHC knew these machines were flawed. It recalled all of them to Tractor Works to change the engine governor and cooling fan. David Baker fitted heavier cast-iron front wheels, hoping to balance the top-heavy tricycles.

At the end of the crop rows, the operator released a rear-axle lock and cranked the steering wheel. This rotated the entire engine, transmission, and drive-wheel assembly in order to turn the Motor Cultivator.

moved the cultivators to straddle the single front wheel. This allowed farmers a "dodging facility" to cultivate without damaging crops within the contours of a row. Still, McKinstry's Sales Department found nothing in this new, fast tractor that it could sell as an advantage over horses. Throughout 1921, Benjamin kept looking. He sent Harry Utley an outline of costs, including overhead, expenses, and income for a farmer operating a 160-acre hog farm. Replacing horses with a Farmall turned feed lands into cash crops. Benjamin calculated a net income of $3,500 versus $3,000 ($49,000 versus $42,000 in 2005 dollars) if farmers with a Fordson still needed six or eight horses for functions Farmalls could serve. Legge circulated this letter around the members of the Executive Council (EC), who already called the Farmall

IHC's tractor engineers made missteps as they invented the next generation of IHC's tractors. It took implement engineer Bert Benjamin to crystallize in the engineers' minds what Benjamin believed farmers wanted.

As soon as David Baker and Ed Johnston shifted the four-cylinder engine off the steering wheels and down onto the frame, stability improved. Baker's first versions still moved as the Motor Cultivator had done, wide-end forward.

"Benjamin's tractor." Benjamin didn't stop there. His time at Ford gave him his next load of ammunition.

While 186 companies claimed to be tractor manufacturers in 1921, and 97 actually assembled one, IHC considered Ford its only challenge. During June and July, Legge held two NWC meetings to consider Fordson and Farmall developments. For the first time, he called various departments on the carpet for foot-dragging. While McKinstry was a Johnston loyalist, Sales cared little for this new machine and never attended field tests to see farmers' reactions to the prototypes. Johnston, criticized for Engineering's slow development of new implements for the Farmall, deflected accusations to the EC for failing to make final design decisions. Utley replied that he'd had little faith in the project initially because McKinstry and Sales showed no interest. McKinstry reiterated that his regional managers felt farmers wouldn't accept anything that forced them to completely re-equip.

Legge listened patiently, waiting until the close of the July 21, 1921, session to deliver his verdict: Farmall implement development would continue at McCormick Works. Tractor Works would expand development to 100 Farmall prototypes for 1922. Seemingly resolved, it broke open again on August 26. Utley pointed out that "the new cultivating devices to go on the Farmall tractor change the machines so materially that [he favored] having a single Farmall tractor brought out and approved before building 100 of these." The EC murmured pleasurably; the farm depression still dampened sales, and Ford's price war had hurt income. But two Farmall champions, Cyrus and Harold McCormick, moved in the shadows, and in late December Legge ordered manufacture of 20 Farmalls with mowers, corn planters, and several other tools.

IHC's EC, its Manufacturing manager, and various Sales department managers were coming around to Benjamin's machine. Only Ed Johnston held out. His motivation was fundamental: The Farmall was not his. It came from an Implements Division engineer who stepped into Tractor Work's territory. Development money for Farmall projects did not go to him. Johnston knew there was little wrong with the idea, but everyone, even those who liked the Farmall, felt it was not quite ready for production.

On February 10, 1922, Alex Legge reconvened the Special Tractor Conference to discuss "the feasibility of bringing out a cheap tractor." The battle with Ford was at full pitch. Benjamin proposed IHC fight back by reducing Farmall power and weight. Chicago sales manager and naysayer J. F. Jones still saw no value in any Farmall, heavy, light, or now cheap. He urged Tractor Works to find ways to "bring the cost of the new 10-20 tractor down as much as possible."

Benjamin persevered. He, Danly, and Baker continued to improve their design, sending drawings to Henry P. Doolittle, IHC's general patent attorney, with paragraphs of text in support, to protect the work. "Time saved in cutting grain with a Farmall," he wrote, "was 12 1/2 percent, or one hour in eight, by turning square corners at full speed . . . avoiding loss of time in turning, and because a smaller headland was required."

IHC's directors unanimously elected Alexander Legge president on June 2. (Harold F. McCormick became chairman of the board. McKinstry became president of IHC of America, the sales arm of IHC.) Legge, empowered by the board and irked by poor prototype quality, wrote Harry Utley on June 23, 1922, chastising Manufacturing for undermining Farmall development. "About the only thing we have demonstrated so far is that we have done a very poor job in putting [Farmall prototypes] together, which suggests that you should strengthen your engineering staff to the extent that you avoid letting out into the field things that have to be rebuilt."

On July 24, Legge and Utley went to Hinsdale to watch Benjamin's light, cheap Farmall. What they saw converted them both. Benjamin's new machine, lighter by 800 pounds, provided three forward speeds but only one reverse. Engineers carried over the differential and turning brake mechanisms, but completely changed the steering and pushed the cultivators far to the front of the tractor. At the next NWC meeting, July 29, the large Farmall died. Addis McKinstry, J. F. Jones, and the rest of the NWC visited Hinsdale throughout early August, each returning to Chicago thoroughly converted to Benjamin's tractor.

Tractor Works engineering documents referred to this as, first, the Combined Tractor-Truck, and later the Cultivating Tractor. Engineering secretary Ed Kimbark first wrote the name Farm-All on November 10, 1919.

IHC's inline four-cylinder engine displaced 220 cubic inches with 3.75-inch bore and 5-inch stroke. IHC assembled perhaps only 25 of these prototypes in 1923.

1923 Farmall Prototype. Neal Stone's rare prototype glistens in afternoon sunlight. By the time IHC's engineers had reached this stage with tractor development, they pretty well had everything right.

1924 Farmall #QC503. This is the earliest-known production Farmall, the third regular-production model assembled. It now is part of Case-IH's historic collection.

At the August 30 meeting, Jones announced "this was the first time that [I] would vote to proceed with the development."

These radical changes in attitudes came about not just because Benjamin's lightweight "cheap" Farmalls impressed his challengers. Early in the summer of 1922, Bert Benjamin's battles with Ed Johnston came to a head. Johnston's failure with the implement-inspired Motor Cultivator contrasted harshly against implement engineer Benjamin's success with his Farmall.

The balance of power in Engineering shifted during the summer and fall of 1922.

In late February 1923, Legge told the EC that Benjamin's tractor came closer than ever to meeting the ideas and answering the objections of everyone who had a say. Benjamin's Farmalls, he explained, would come from Johnston's Engineering Department, made by hand, as previous

The Farmall sat 86 inches wide, overall, and used a 74-inch rear track width and a narrow 10-inch width for the front wheels to slip between crop rows.

models were done. His implication was clear. Legge and the group unanimously approved assembling 25 of the tractors, with "no radical departures to be made in design." A chastened Ed Johnston promised that his Engineering Department would begin production in mid-May and be completed by early September.

Johnston and Legge calculated mass-production costs and found it was nearly the same as the 10-20 Gear Drive. Legge ordered Johnston to stop experimenting with other final-drive systems so they could firmly set Farmall specifications. Legge needed Sales and Manufacturing to determine prices and promotional methods of the tractor, which got its "Farmall" trademark registration in Washington on July 17, 1923.

IHC began assembling "production" Farmalls on December 26, 1923. The Executive Committee had authorized only 200 pilot production models for 1924 and a small crew completed the job by late May, sometimes assembling as many as two a day by hand.

Its appearance startled everyone who saw it. Tractor design and engineering had evolved in the first quarter century from giant steam traction engines to this, which one Texan called, "Homely as the devil," but if you don't want to buy one, you'd better stay off the seat."

The Farmall measures 123 inches overall in length on an 85-inch wheelbase. It stands 67 inches to the top of the steering wheel. Basic Farmall models weighed 3,825 pounds.

IHC submitted its Farmall to the University of Nebraska for testing in mid-September 1925. Running for 39 hours with no repairs or adjustments, test lab staff observed a maximum drawbar pull of 2,727 pounds, 12.7 horsepower while it measured a peak of 20.1 horsepower on the belt pulley.

A small crew assembled these first 200 tractors on sawhorses in a corner of the 10-20 Gear-Drive plant. Each tractor then got a three-hour engine break-in out in the field behind the plant.

Johnston completed 26 models long before his deadline, and by August 9 they were working. Four remained at Hinsdale, one went to Cyrus Jr.'s farm in Wheaton, and another to Utley's in Downers Grove. Johnston shipped 13 to branch houses in Georgia, Mississippi, Tennessee, Texas, and Wisconsin for evaluation by farmers. Almost without exception the results were good and the tractors held together. Reports to Legge said it had "plenty of power," and "splendid ground-gripping qualities." Benjamin interviewed operators on the Durham Farm in Wayne, Illinois, where one hand told him, "In cultivating young corn four or five inches high, we can do a better job with the Farmall because it handles so easy that practically all our time [is] utilized watching the cultivator shovels. With a two-row cultivator pulled by three horses, we pay so much attention to the horses that we cannot do a good job cultivating."

On October 9, McKinstry recommended Utley's department produce 100 Farmalls for 1924. Legge and Johnston, encouraged by the reaction among southern testers for the Farmall, advocated higher production even though they worried about its appearance. While it did "at least . . . look like a tractor," it was not attractive. A cotton farmer from Texas who owned two put it concisely: "It's homely as the devil, but if you don't want to buy one you better stay off the seat."

1924 Farmall Regular. Implement engineer Bert Benjamin carefully followed the progress and work experience of his first 200 Farmalls. While the initial dozen or so went to Texas, others scattered around the country.

FARMALL

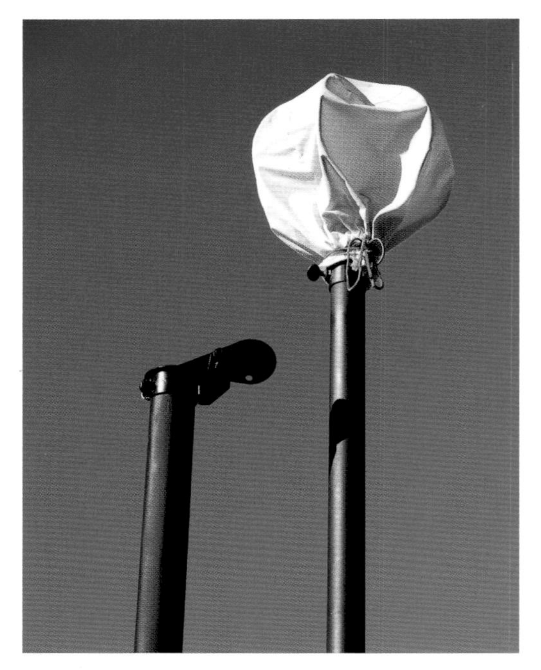

The early production air cleaner element was a flannel-covered banded bag. Its vulnerability under low-hanging trees made it one of the Farmall elements to be changed as Bert Benjamin and Tractor Works engineers got feedback from owners and operators around the country.

Tractors authorized for 1924 production gained improvements that included enlarging rear-axle diameter; increasing drawbar fastening size; fabricating starting gears of steel; making main frames of rolled tubing; and strengthening bull-gear housings. Tractor Works also enlarged kerosene fuel tanks to 13 gallons.

Selling for $825 (about $11,550 in 2005 dollars), IHC lost substantial sums because these tractors were hand built. The EC accepted this as introductory and promotional expenses. In late February 1924, the NWC and EC, still moving cautiously, ordered 200 Farmalls built to expand testing. On leap year February 29, Tractor Works shipped QC-501 to Taft, Texas, where a farmer named Roy Murphy got to use it. Soon after, Benjamin sent Murphy the first "skeleton" wheels, open metal frameworks soon known as Texas wheels.

Benjamin spent two weeks with QC-501 in Taft. Murphy asked him for a four-row cultivator and, working with a local blacksmith, Benjamin fabricated a prototype. Murphy called in his neighbors to watch as he covered 100 acres in 14 and one-half hours compared to 18 acres the previous 10-hour day with his original two-row cultivator. Benjamin took plans to McCormick Works for production for south Texas cotton farming and the NWC approved production of 50 units for further testing.

By September, Tractor Works had turned out 205 of Benjamin's tractors and improved, strengthened, modified, or revised 16 more features ranging

Right
Ed Johnston, Bert Benjamin and IHC's patent lawyers worked frantically to protect all the innovations they had made on this new row-crop tractor. The "patent pending" notification was nearly as large as the tractor's model name, Farmall.

Opposite
As early operators complained about noise, Farmall engineers improved the engine exhaust configuration. They replaced the tall open pipe with a muffler.

McCORMICK-DEERING
FARMALL
PATENT
PENDING

1930 Farmall NT. Tractor Works introduced a Narrow Tread model for "export" on May 5, 1927, mainly to Argentina. IHC had acquired Moline Plow Company's Rock Island, Illinois, tractor factory in 1924. This allowed IHC to accelerate production and begin special models when production stated there in June 1926.

from simplifying the steering gear, to fitting a muffler in place of the overhead exhaust pipe, to simplifying the gear shift and further strengthening the transmission case. When J. F. Jones, the reluctant convert now proposing manufacture of 300 units, wondered aloud one last time about the harm the Farmall might inflict on 10-20 Gear Drive sales, it was Ed Johnston who answered him.

"There has been a constant cry for a tractor to meet Ford's. It is impossible to meet Ford on price; therefore, we have to produce something of greater utility to justify our price. The Farmall will do this. If the Farmall is the tractor that will kill the 10-20, it would be far better if we ourselves kill it." With the last critic at peace, the NWC and EC agreed to manufacture 250 units. Dealers

This original, unrestored example shows little use and extensive care during its lifetime. IHC began using its E4A magnetos on Farmalls after 1926.

in some branches were told to not promote the new machine but instead to push 10-20s and 15-30s, delivering Farmalls only when buyers insisted. Sales told branches in other areas they could not get one at all. In the cotton south, McKinstry's staff encouraged Sales.

Legge put this strategy in perspective: "The Farmall was going into the cotton states," he said, "and if we did not push it elsewhere there was little danger of hurting 10-20 sales . . . , on the other hand, it would pick up business that we had never had." During the EC meeting on March 21, 1925, Benjamin sparred with J. F. Jones over another cost analysis. He'd examined cotton production with mules at $110 a bale against $83 a bale with the Farmall ($1,540 against $1,162, adjusted to inflation). "You see that operating by Farmall . . . makes it possible to save the crop for the United States instead of losing it to Egypt, southern Russia, India, the Argentine . . . , producing [there] with cheap labor at $95 a bale" ($1,330 in 2005 dollars). In April, J. F. Jones went to Texas to see for himself.

Outside of San Angelo, Jones watched a demonstration with P. Y. Timmons, manager of tractor sales; Jim Ryan, Houston branch manager; Joe Foley, Dallas' manager; Guy Fisk, Amarillo's branch manager; several implement engineers; and Bert Benjamin. As the afternoon wound down, the group began discussing the Farmall. Timmons reflected J. F. Jones' lingering concern that it might adversely affect 10-20 sales. Fisk, Foley, and Ryan agreed they "had little use for [10-20s] in cotton country. They were not adaptable to all row crop operations."

Farmall production reached 35 a day as IHC introduced these narrow-tread models. Bert Benjamin and Ed Johnston quickly began to devise and develop variations for a wide variety of crops.

Narrow-tread (NT) models received an offset rear hub. This collapsed rear tread width from the standard 74 inches down to about 63. Documents in archives suggest an even narrower version provided 57 inches but this appears more commonly on Fairway tractor specifications.

Jim Ryan didn't hesitate: "If you don't adopt it for production, we'll organize a company in Houston and build it down here . . . ." Jones, parroting Ed Johnston's comments to him eight months before, announced that "if anyone was going to build a tractor that would affect the sale of the 10-20, let's do it ourselves." An official photo taken at the demonstration showed Jones near the planters towed behind the tractor. Benjamin, at the engine, stood a long way away.

ON MAY 5, JONES STRESSED THE NEED to have 1,000 Farmalls ready by the end of October. Within a month, the number increased to 1,500 and at the end of July, McKinstry added

Production in 1930 reached the peak, at 42,093 Farmalls completed at the Rock Island plant. On April 12, IHC celebrated completion of the 100,000th Farmall.

another 1,000 to the plan for 1926. Good crop harvests in 1924 put ambition in farmers' eyes in the winter, creating more interest in the Farmall. In September, the EC ordered Tractor Works to solve the two chronic Farmall complaints about broken studs on the differential and troubles with the steering.

On March 19, 1926, Utley reported to the EC that Tractor Works produced eight Farmalls a day, having met outstanding orders for 1,708 tractors. IHC priced them $100 higher (about $1,300 in 2005 dollars) than 10-20s to emphasize their greater potential. The build order climbed to 2,954 and Utley expected daily production to reach 15 by July 1.

FAIRWAY

Typically, Fairway models had small conical lugs bolted into the steel wheel rims. These provided traction without tearing up the lawns and golf courses.

The 3.75 x 5-inch incline four-cylinder engine did well enough in its first years. But it didn't take long before Ed Johnston and his colleagues began talking about increasing power again.

McCORMICK-DEERING
FAIRWAY

IHC could make this jump because in June 1925, it opened the Farmall Works. It acquired this facility in 1924 from Moline Plow Works which had since been IHC's tractor plant in Rock Island, Illinois. Legge told McKinstry to "make no larger capital expenditures at Tractor Works to increase production on the 10-20, but proceed at once with the development of the Rock Island plant layout for floor space for an output of one hundred a day." By November 5, 1926, near the end of another record crop harvest, Rock Island settled at 20 to 25 per day. Threshing had begun throughout the Midwest when hard rains and high winds hit in storms lasting days. Harvesting and threshing stopped, fields flooded, and unharvested grain shocks and bundled stacks sprouted. Some states' grain harvests were entirely ruined. Crop prices slipped rather than rising from destroyed supplies. Predictably, tractor sales fell. While U.S. makers produced 178,074 tractors in 1926, they sold only 122,940 (46,441 as exports); 4,430 U.S. sales were Farmalls. McKinstry hoped 1927 would reach 6,600 while district sales managers believed 7,500 was more realistic. Legge took a leap of faith and he guessed right: Rock Island manufactured 9,502 throughout 1927.

Tractor Works introduced a narrow-tread model for "export" on May 5, 1927, just as Rock Island production reached 35 a day. Production for 1928 reached 24,899 Farmalls: 35,517 in 1929, then 42,093 in 1930 (with daily totals reaching 200 on January 27, 1930). As the U.S. Census released figures counting 920,378 tractors on farms, IHC celebrated on April 12 when Rock Island's Farmall Works completed manufacture of the 100,000th copy of Benjamin's tractor. (Within two months, the corporation celebrated again as the 200,000th Johnston 10-20 Gear Drive, with Cyrus McCormick Jr. at the wheel, rolled out of Tractor Works.) But in 1931, the agricultural depression, made worse by a long drought in 1930 that destroyed crops, caught up with what began on Wall Street in October 1929. Farmall sales collapsed by two-thirds to 14,093 for the year. Competition fell away too; where there had been 186 registered tractor makers in 1921, in 1930, there remained 33. IHC staked its claim as the dominant tractor maker in North America. However, other farm equipment makers were not standing by idly.

1927 Farmall Fairway. Fairway models appeared early in the Farmall series. These tractors went out on 16-inch-wide rear steel wheels and 8-inch-wide fronts.

Except for the wheels, IHC changed no other specification to produce its Fairway models. Because Fairways and Regulars came off the same assembly line, IHC records never separated the golf course models out for a separate count.

Benjamin and Johnston, ever perceptive, recognized the trends for both smaller and larger machines. As the January 9, 1930, issue of *Farm Implement News* reported, "Three-plow standard type tractors have added . . . to their stature . . . because the Great Plains wheat farmers discovered that they could handle another 100 acres or so with wheatland disk plows, duck-foot cultivators, rotary rod weeders, and combines, if their three-plow tractors only had enough more power to pull an extra foot or two of disk or to carry the combine up slopes." In early 1930, mention of projects around Gas Power Engineering Department began to appear in NWC notes and EC minutes: committee members referred to them as the "Increased Power Program." But they weren't describing changes forced onto 10-year-old 15-30s and 10-20s.

Farmalls were very successful tractors. Between 1924 and 1932, IHC produced 134,647 of the Regular and Fairway models in standard and narrow tread variations. This machine defined power farming and for decades afterward, every row crop tractor was a "Farmall."

Farmall Becomes a Big Family

Chapter 4

1930–1935

"Progress," Ed Johnston told IHC's Executive Council, "has put our competitors in a position to increase the horsepower for the size of engine and to improve the fuel consumption. We are suffering in the trade." He urged the EC to adopt a program to produce a more powerful Farmall, and even a smaller one. EC members promptly approved three sizes, counting the current Farmall and the proposed "increased-power Farmall" as one. The second was an intermediate Farmall, using an increased-power engine for the 10-20 tractor. Third was a large Farmall to use the increased-horsepower 15-30 tractor engine.

While the power increases came from a new head, intake manifold, and piston design without changing bore or stroke, to Johnston, the term "increased power" meant "improved tractor." He installed a water pump with a more effective thermostatic control to improve cooling. He strengthened frames because some Industrial Model 20s in Europe had broken. In August, he created a wide tread for the Farmall from 10-20 parts for crops around San Francisco. The Gas Power Engineering Department's experimental engineers turned out

1936 F-20. Ed Johnston created a non-adjustable wide front-end for both F-20 and F-30 models in the spring of 1932. This coincided with the beginning of production for both models.

FARMALL

"Progress," Ed Johnston explained to IHC's executive committee, had allowed their competitors to catch up with the Farmall. This was his secret weapon to move back into the lead. Johnston got this Increased Power Farmall photographed on May 3, 1930.

new engines and strengthened unit frames on IHC's wheel tractors. Johnston's dictum to improve the tractors became the goal throughout GPED.

In New York, while stock market prices still fluctuated wildly, President Herbert Hoover asked Alex Legge to help stabilize farmer market prices. Legge had resigned from IHC's presidency to head the Federal Farm Board in June 1929. He remained on IHC's board, however, and occasionally returned to Chicago for meetings that particularly interested him, especially when his efforts in Washington made little progress.

On December 1, 1930, Legge redefined the experimental Farmalls in terms farmers understood. The Increased Power model would handle two plows; the Intermediate, based on the improved 10-20, would run three plows; and GPED had designed the large 15-30-derived Farmall for four. Then he wondered if this incremental power increase was large enough. David Baker idly suggested fitting four-speed transmissions into the large and intermediate models and simply

This F-20 was the middle prong of Ed Johnston's three-way attack on IHC's competitors. Using an Increased Power engine from the 10-20, this became the new "intermediate" Farmall.

At the low end of the Farmall range, Johnston and Gas Power Engineering created the F-12 intended for the small operation farmer converting from horses or mules. This was one of the first 25 assembled, showing its integral 16-inch plow. The date was November 15, 1932.

dropping the two-plow original "regular" model. A newcomer to New Works Committee meetings, John L. "Mac" McCaffrey, IHC's 38-year-old assistant manager of Domestic Sales, disagreed. He reported that his boss, Maurice F. Holahan, manager of Domestic Sales, "felt we should put out the Increased Power Farmalls only at this time."

Two weeks later, the Naming Committee gave final designations to the Increased Power models, referring to the two-plow model as the F-20, the Intermediate as the F-30, and the Large Farmall as the F-40. Alex Legge held a special conference on IHC's tractor development and business in Phoenix, March 12–14, 1931. He had left Hoover's Federal Farm Board on March 5, frustrated by his inabilities to help farmers and turn around the world economy. He returned as president of IHC and immediately got to work where he knew he made a difference.

"Up to 1914, the tendency of tractor design has been entirely toward larger machines of greater horsepower," he said. Recognizing this strategy was not economical for the average

Bert Benjamin and Ed Johnston worked on a variety of systems to raise and lower implements. This compressed-air system never went into production but was photographed on November 13, 1934.

Engine output increased because Ed Johnston's engineers designed and tested new cylinder heads, intake manifolds, and pistons. He added a water pump, which greatly improved engine cooling and lubricant life.

1930 Improved Power Farmall Prototype. Ken Holmstrom's prototype shows subtle features that distinguish it from late-production Regulars. The canted front wheels are the first giveaway. Internally, the improvement in power came from new engine pieces.

farmer, IHC frequently "went right about-face to the small tractor and in 1914 we made our first, the one-cylinder 8-16 Mogul. The same economic conditions govern us today, and we must not make our tractors too heavy, too high in cost, and too expensive in operation."

Bert Benjamin proposed returning to the Two Tractor Plan. He suggested that a 24-horsepower Increased Power Farmall, along with the Intermediate Farmall, would take care of 90 percent of IHC's current business. The other 10 percent came from California, where IHC's new crawlers would fill their need. The EC approved large-scale production of the F-30 Intermediate-sized Farmall and the Increased Power F-20. Two matters remained.

IHC relied on its E4A magneto from model introduction in 1942 for another few years before replacing it with the F4.

David Baker reported progress in engineering the high-compression, heavy-fuels projects, particularly with the Hill diesel engine from Michigan. One prototype ran at Phoenix with pump-wear problems that they solved on the spot. Baker mused that "if this Hill diesel engine came through with satisfactory performances, our troubles on the Increased Power 10-20s and 15-30s would be behind us, as those two tractors would [use] the Hill engine." Legge urged Baker to get to work.

On July 14, board chairman Cyrus McCormick Jr. signed off on improvements to the "Regular" Farmall, including raising output by 3 horsepower, adding a four-speed transmission, and enclosing the steering gear. Two months later, on September 26, 1931, Johnston asked Baker to lay out and design a new one-plow Farmall, an F-10, with their modified unit-frame at the rear for the engine, transmission, and running gear while extension rails, mounted onto the unit-frame, supported the front axle.

In January 1932, the NWC dealt with variations on old themes, modifying the regular narrow Farmall as a new Fairway tractor by replacing the front wheels that often cut into the bunkers,

These Schebler carburetors lasted only through the prototype phases. Once the tractor went into production as the F-20, Gas Power Engineering began using 1.25-inch Zenith K5 models.

Road rims let Ken Holmstrom enjoy operating his unusual machine in parades and shows. Nearly every dimension on this prototype falls between production F20s and F30 models

Ed Johnston used these prototypes to test and develop the cambered front wheels. These, and the "duck-bill" steering column that topped them, went into production in 1932.

with a wide front axle. Ed Johnston created a similar configuration as a wide-tread front axle for the F-30 narrow-tread tractor in late February, as a no-additional-charge option. He used wheels off the 10-20 and the wide-front versions of the Farmall F-30, the W-30 tractor, but the F-30 wheels would not fit an F-20. By April, his engineers devised a wide front axle for the F-20 that also would fit the Regular Farmall.

David Baker continued work on the F-10 one-plow tractor. Tractor Works released the first of these smaller semi-unit-frame models as the F-12. Competition nipped at the Farmall's heels. The Regular's output dipped to 3,080 for 1932, though IHC tractor prospects were bolstered by production of 2,500 F-20s as well as 1,500 F-12s.

Opposite
1935 F20 with All-Weather Covering. This was the closest thing a Midwest farmer could get to an enclosed cab for cold winter work. "The Heat Houser," manufactured by a tent maker in Fort Dodge, Iowa, did a fair job of directing engine heat back to the operator.

THE TREND TOWARD INFLATABLE RUBBER TIRES caught up with IHC. On August 1, 1932, GPED's Leonard Sperry urged the NWC to keep up with competition. "Because of other tractor manufacturers, we [must] offer low pressure tires," he said. "We can purchase wheels from French & Hecht which are built to fit our tractors and are sold to our dealers. The tires have inner tubes and carry about 12 pounds pressure, not the zero pressure [solid rubber] tires tested in Florida. We will ultimately furnish Goodyear tires when they are ready." At year-end, as the NWC cleaned up loose ends, they heard from Leonard Sperry again.

WEAVER & LINGG
SALES & SERVICE
Heat-Houser

Michigan Farmall collector John Wagner recalled that this system was not perfect. "Going upwind, your feet and legs roasted," he explained, "and downwind your backside froze."

"The unexpected has become the accepted," he observed. "It is entirely possible that pneumatic tires may be developed to meet many agricultural operations as they are now meeting industrial tractor needs. Allis-Chalmers are advertising pneumatic tires on farm tractors. There is a possibility that these tires may cut into crawler tractors sales: Caterpillar is experimenting with pneumatic tires on wheel tractors it has purchased."

The NWC, aware of Caterpillar's growing role in influencing IHC product development, approved low-pressure pneumatics for the increased-power 10-20s, W-30s, and the Farmall Regular as well as the F-12 tractors. They offered F-12s as orchard, industrial, and fairway versions.

1935 F-30HV. This was IHC's first cane high-clearance F-30, serial number FB7262CNW. The suffix represents Cane tractor, Narrow rear tread, Wide front end.

Sperry recommended providing a fourth, much higher speed gear, determining that 10 miles per hour now seemed sensible.

In early 1933, the NWC addressed ongoing problems large and small: magnetos and impulse starter couplings for four- and six-cylinder Farmalls; a clutch-release hitch for Farmall tractors that disengaged if the plow hit something; low-pressure pneumatic tires for the F-12; worm-steering gears for the F-20 and F-30; kerosene engines for the F-12; continued pressure on Sperry and GPED for diesel engines; electric starters; relocated air cleaners; new engine crankcases to provide better lubrication to the bottom end and tops of the F-20

engines; corresponding widening of the tractor frame with hood and fuel tank to match; elimination of engine side doors to clean up appearances of the same tractors; and prototype clutches for the F-12 produced in-house to replace outsourced units.

"The committee," Ed Kimbark read in a letter from Harold McCormick during a June 20 NWC meeting, "are impressed with the advantages of the construction used in the F-12 Farmall, compared to the Regular, the F-20 and F-30. Designing this form of chassis, having high wheels and one-chamber gear case should be combined with designing modern, higher-speed, four-cylinder engines for tractors of the two-plow and three-plow sizes."

Leonard Sperry went pale. Was this an order to redo the entire line? His portion of GPED was overextended as it was. Ed Johnston scrambled. His resources also were stretched thin. The

The F-30 was IHC's biggest tractor at the time, stretching 147 inches long. These cane high clearance models reached nearly 100 inches in the air.

Opposite
P.Y. Timmons, IHC's power-farming-equipment sales manager, alerted IHC's management that farmers had reintroduced sugar cane as a crop in Louisiana. By late 1933, Ed Johnston's engineers were at work to create a useful machine.

FARMALL

Engineering had to design and fabricate a new front axle in order to make these tractors. The standard F-30 front axle was an inflexible casting. It could not be arched to provide higher clearance.

The F-30 engine displaced 284 cubic inches, compared with the F-20s (and Regular's) 220 cubic inches. Cylinder bore grew to 4.25 inches from 3.75 while stroke remained at 5 inches.

By 1935, all F-30s used 1.25-inch Zenith K5 carburetors. The engines developed 20.3 horsepower on the drawbar and 30.3 off the pulley or PTO.

F-12 meant a lot to him. It was IHC's first mass-production application of the unitized tub he had created with a large prototype Mogul 20-40 back in 1914. He had furthered the benchmark that Bert Benjamin had established with his Farmall.

But he had no money, no personnel, and no time to develop new tractors, even ones based on his own idea. The high wheel design presented obstacles to attaching existing Farmall implements. While IHC had reduced tractor-manufacturing costs by using a unit frame, the F-12-type case reduced clearance for cultivating, and brought wheel rims and their dust closer to the driver. The changes required to existing implements outweighed the advantages. Johnston voted to avoid killing his staff or turning out another hurried project.

The four-plow tractor idea returned in discussions on September 11, 1933. Based on semi-unitized F-12s with speeds ranging from 2 to 20 miles per hour, the NWC now wanted this with the diesel engine and it pushed a rapid development program for Cane Cultivator-and-Plow tractors based on F-30-N narrow-tread models. Sugar cane had come back in Louisiana and planters needed tractors with high ground clearance.

Sperry agreed the F-30-N Farmall offered the "advantage [of] straddling the bed in cultivating instead of running between the beds as the 10-20 must do," but the narrow-front axle was cast and couldn't be arched without redesign and new manufacture. Suddenly, everything Johnston and Sperry supervised felt rushed. They never had enough time or personnel to do the work, although they always got enough money to do the projects thoroughly. Their effort and IHC's product inventory expanded geometrically. Sugar-cane versions of F-30s rose high off the

Michigan farmer John Wagner raises seed corn, another crop for which a high-clearance tractor was useful. However this historic piece has retired now, and emerges for shows and photography sessions.

ground; Orchard, Industrial, and W-series standard-tread variations of the Farmall F-12 were tested and went into production. A dozen times in the last six weeks of 1933, Ed Johnston repeated, "The Gas Power Engineering Department could not do much on it at this time, without an increase in the engineering force, for the reason there are so many developments under way." It became his litany. The pace of work was nearly unsustainable.

Then, for a few brief days, all work halted. Alex Legge, shrewd champion of tractor development and the man who had led IHC through tractor sales wars against Henry Ford and the relentless onslaught of competitors real and would-be, died of a heart attack in his garden as he pruned his lilacs on Sunday morning, December 3, 1933. After the funeral on Wednesday, December 6, the McCormicks and the board of directors elected Addis McKinstry president. He would serve only until New Year's 1935, before retiring.

During one of McKinstry's first NWC meetings on January 15, 1934, the Naming Committee relabeled the improved gas-engined wide-front 22-36 tractor as the WA-40. IHC's first diesels became the WD-40. McKinstry signed orders to produce 20 of each per day for two years. GPED completed WD-40 detail drawings on March 1, sent them to Manufacturing, and

IHC assembled the first F-20 in early January 1932. By the time the company stopped producing these tractors more than 154,000 had gone out the doors at Rock Island Farmall Works.

The inline four developed a peak of 15.4 drawbar horsepower and 23.1 horsepower using distillate fuel during Nebraska's tractor lab tests. The best performance a Regular recorded at Nebraska a decade earlier was 12.7 drawbar and 20.1 belt-pulley horsepower.

the first development models, seven WAs and three WDs, rolled out of GPED on May 1. These meetings marked the earliest appearances of the newest McCormick, Fowler, to the EC and the NWC. Born in 1898, he was Cyrus Hall McCormick's grandson. After graduating from Princeton at 23, he drifted through several interests, psychology (inspired by his mother's fascination with Dr. Carl Jung), music, and accounting. He ran a small business until 1928 when Alex Legge suggested it was about time he came to "The Company." Fowler started in the apprentice program. He spent five years learning manufacturing, engineering, and sales. By 1933 he was assistant sales manager, and in 1934, McKinstry named Fowler head of foreign sales.

In late May 1934, IHC found itself needing to repair its reputation as had happened several years earlier with some of its crawlers. Now, a combination of design, manufacturing, and field service problems produced an ill-fitting air filter for the diesel engines and the gas F-12s.

Climate-related, it appeared only where drought conditions brought on intense dust. Johnston explained: "The numerous complaints of excessive wear of the F-12 engines are largely due to dirt entering through inefficient air cleaners. The design of the original made it extremely difficult to assemble the cleaning element uniformly into the cleaner. It results in excessive wear of pistons, sleeves, rings, bearings, and crankshafts which in turn results in excessive oil and fuel consumption and loss of power." (Sperry defined acceptable oil use as one quart per 10-hour field day after 300 hours of use; these tractors used 1 gallon a day.)

IHC's policy was to make it right by replacing faulty air cleaners with new ones, making engines airtight by, if necessary, thoroughly overhauling the engine and replacing worn parts. GPED designed new pistons with four rings instead of three. One source of the dirt was residual sand and metallic chips from Milwaukee and Tractor Works castings. McKinstry ordered them to install a filtering system for the lubricating oil they used to run in engines prior to installation. IHC had these systems in operation at the Fort Wayne gas engine plant and at Farmall Works.

"I think each individual identified with this undertaking," McKinstry said, "must be impressed with the amount of money involved." McKinstry kept records but released no totals

The F-20 was the successor to IHC's groundbreaking row-crop Farmall. With internal engine changes and new exhaust manifolds, engineers initially pushed 20 percent more power out of the engine. Test results using various other fuels often showed more of an increase than that.

1936 F-12. After assembling 25 pre-production versions of this tractor, Rock Island Works got down to business and series manufacture started on January 11, 1933. IHC continued to produce these models into 1938.

for this repair; it would have been premature. He had learned how to manage from Legge, and the new president quickly found his voice, using it effectively to satisfy customers and to ask his managers to work more carefully and more wisely. It worked.

The pressure from domestic sales to provide a tractor for every farm and crop strained GPED by mid-1934. Johnston's durability and stamina earned him promotion to vice president of Engineering. Yet even now, as a corporate officer, he couldn't slow the flood of new work. Two new projects lined up behind each one completed. Engineers returned from one test trip, filed reports, and left for another. The W-40s would start production even as 1934's Midwest drought slowed demand for all IHC products. Johnston and Sperry hoped for an opportunity to catch up.

When the costs for working at this pace came due, the price was high. At noon on June 27, 1934, McKinstry and Johnston shut down the 12-series tractor production line, halting manufacture completely. It was the only way to get parts changed before 12-series tractors left the plants. Repairs cost much more in the field. This delay permitted outside makers of new air filters and elements to deliver adequate supplies, so Tractor Works could remedy the problem before shipment. Production resumed on July 9.

The entire program, including parts, repairs, overhauls, and F-12 plant shutdown, cost IHC $750,000 (close to $9 million in 2005 dollars). In four regions, Central, Southern, Southeast, and East, 19 branches needed help; the 36 others around the country did not. The Service

This 1936 model uses IHC's own 113-cubic-inch four-cylinder engine. The first 2,500 or so that the company assembled used Waukesha engines of 3- inch bore and 4-inch stroke. IHC's engine would have identical specifications.

Department trained sales agents and sent them out to make repairs. Manufacturing estimated that perhaps 10,000 diesel, F-12, or crawler tractors needed service, ranging from simply tightening or replacing air filter canisters and elements, to full top-to-bottom engine rebuilds, transmission repairs, and, in the case of crawlers, track replacements.

Ed Johnston had harped about keeping dirt out and oil in. IHC's sales organization, hungry for products and concerned with manufacturing costs, continually postponed his efforts to make tighter machines. Johnston had argued for pressure lubrication; he got that with two of the company's crawlers. Yet, even as this unprecedented "recall" continued, Sales argued that "the appropriateness of pressure lubrication on farm tractors as compared to splash lubrication had not yet been fully demonstrated."

In early August, David Baker sent a new wide-front four-plow tractor, the CW-40, to Hinsdale. Johnston, intent on avoiding recent problems with tractors released too quickly, asked for another year for testing and development. McKinstry reminded him that this third series "W" model was the 22-36 replacement. The C-version only incorporated the latest seals and air filter. Test harder, McKinstry said. He refused to delay production.

GPED staged final sign-off tests of the third preproduction W-40 series in early October 1934, in Phoenix. Johnston chose the desert there to guarantee challenging conditions. The only problems came in the transmission. While it never failed in tests, the heat and dust taxed it. In Johnston's ideal world, GPED wanted to upgrade the transmission before the tractor grew from the 15-30 to 22-36. No one envisioned the power of the diesel. Larger gears wouldn't fit; designing a new case and testing a transmission would add two to three more years. Again, McKinstry refused to delay introduction. Disputes such as these between Sales and Engineering set the stage for a drama that would play out over the next half century.

Tested at University of Nebraska in May 1933, the F-12 developed 10.1 horsepower at the drawbar and 14.6 off the pulley or PTO. It weighed 2,700 pounds, compared with 3,950 pounds for the F-20 and 5,300 pounds for the F-30.

1936 F-20. Little rain made for hard, dry soil and plenty of bean stubble for the two-bottom Little Genius 12-plows and F-20 to work through. The F-20's four-speed transmission and extra power advantage over the first generation Regular Farmall gave farmers plenty of strength to get through tough conditions.

ED JOHNSTON'S GPED WAS OVERWORKED; however, Bert Benjamin had not been resting on his accomplishments. In early October 1934, at Hinsdale, he showed NWC and Executive officers "a new means of attaching implements to the F-12 tractor," demonstrating both a No. 90 plow and a middle-buster. Predictably, the Sales staff and McCaffrey, most vocally, wanted all of it immediately, and available universally.

Benjamin and Sperry had resolved to introduce this new hitch for plowing after farmers had completed harvest for 1935. They could deliver F-12s earlier because hitch modifications were small. Implements were the problem; the list to be offered grew like weeds. With tractor production at 2,000 per month and the 1935 fall harvest 10 months away, tractors and enough implements had to be in dealers by July. In late November, Sales convinced everyone that tractors without implements were preferable to new implements without a tractor. This strategy gave them time to advertise and farmers time to anticipate.

With McKinstry's admonition about no delays searing their ears, Johnston and Sperry wrote to McCaffrey. Having watched him for nearly two years by now, they recognized an enthusiast in the supersalesman. They advised him that their large six-cylinder diesel "with recent modifications could be run safely at a speed of 1,500 to 1,600 rpm, and could be depended upon to develop the horsepower required. The Engineering Department could turn over specifications for this engine in two months."

In John McCaffrey, they imagined the preservation of their world, if not a renaissance for engineering. They hoped to build for him high-quality machines, tested thoroughly, and put into production when they were ready. But as Addis McKinstry had learned things from Alex Legge, John McCaffrey was learning from McKinstry, and McCaffrey was not the last one who would disappoint the engineers in the decades to come.

Problems and Successes Continue

Chapter 5

1935–1944

The inaugural New Work Committee meeting of 1935 presented a case of déjà vu: IHC had trouble again with piston ring wear, particularly oil control rings, just months after field service technicians had completed the improved parts' "First Change." IHC had launched F-12 production using a Waukesha-built engine designed by Fuller & Johnson Manufacturing Company because GPED's own was not yet ready. Once IHC began using its own engines, problems quickly developed.

The kerosene manifold didn't heat the fuel well enough to vaporize it completely. Liquid flowed past piston rings, diluting lubricating oil. Worse, Leonard Sperry had sampled manifolds and found that in 40 percent of them, the cored hole missed the carburetor. Ed Johnston uncovered another cause that was out of IHC's control: bad fuel. Distributors, stuck with poor-quality heavy oil, added high-test gasoline to it so the mixture would ignite. This blend broke down and combustion did not consume some of the elements, leaving damaging deposits. It was a geographic consideration. There were no problems in Texas, yet

1938 F-20. Perhaps it's the narrow wide-front end that's captivated them but the herd came closer to inspect the machine. These weren't that rare, however. IHC produced more than 154,000 of its F-20 models.

FARMALL

nearly every tractor sold in Little Rock and more than half from Memphis required an overhaul and a new crankshaft.

"Should we stop production of the kerosene-engine F-12 tractors?" Leonard Sperry asked. John McCaffrey voted "no" on interrupting production. "We are practically current on orders," he said. "Three out of four F-12s are kerosene tractors. Better we face having to send changes later than to not ship tractors again." Johnston and Sperry hurriedly created new manifolds and revised carburetors with a smaller venturi and a new fuel nozzle. They covered both pieces with a heat shield to deflect the fan blast that overcooled the intake manifold. They fitted these "Second Change" kits to tractors still in the assembly line.

In early February, GPED's managers examined the new hydraulic power-lift built into the F-12 Farmall. They mounted the pump close to and driven by the power take-off shaft. They

IHC didn't introduce its diesel wheel tractors to the public until the Model WD-40 of the spring of 1935. However, engineers tested a variety of diesels for years before. This 1930 prototype inline four-cylinder Hill diesel used a low and high compression chamber to facilitate starting.

In 1931, Ed Johnston's efforts to increase the power for IHC's Farmalls yielded this prototype diesel tractor. It wasn't until a New Work Committee meeting in January 1934 that IHC's Naming Committee christened these wide-front diesel-powered models as the WD-40s. Addis McKinstry's authorized production of 20 a day for two years.

had examined an implement lift from a John Deere Model A tractor. Deere's system, designed by its own version of Bert Benjamin's, an engineer named Theo Brown, provided farmers with hydraulic lift, but Brown's design used gravity drop. IHC mounted their lift on the rear axle, giving their patent lawyers worries. GPED engineers proposed relocating the lift ahead of the rear axle, but attorneys doubted "the interpretation 'forward of the rear axle' differed enough from Deere's placement 'at the rear of the tractor.' "

Meanwhile, in the Tractor Works' backyard, EC members watched the new prototype F-20X, built along F-12 lines. This was Johnston's newest application of the semi-unit frame design of his 1914 Mogul 20-40.

By late March, Sperry presented Sales with a new advantage that F-12s offered over John Deere models: GPED's power implement lift put pressure on the implement to set it into the ground after it was dropped. It was ready for testing. Another F-12 invention eventually spread to the rest of IHC tractor lines, the Quick Detachable "QD" Draw Bar, proposed in October 1934. By April 1935, GPED had working prototypes. The interrelationship between Engineering, Manufacturing, and Sales created a new production plan. Manufacturing accelerated the pace of all tractor lines in late April 1935 to develop a steady schedule, instead of letting market responses create peaks of productivity or valleys of plant inactivity. This stabilized the labor population as well. But now, stockpiling tractors created another problem.

"When tractors stand exposed to the weather more than 30 days," A. W. Seacord of Domestic Sales warned NWC members, "it is usually necessary to repaint the tractor at a cost

1937 Model WD-40 diesel. IHC manufactured the first WD-40 on April 16, 1935. By 1940, prices for these 7,500-pound machines reached $2,516.50 on pneumatic rubber tires.

of about $5.00 each [about $60 in 2005]. This occurs whether the exposure is in storage at factory yards or outside at branches or with dealers." GPED proposed using new "synthetic paints which must be sprayed and which require a somewhat different provision in drying ovens . . . [and] will allow an exposure of four to six months. If the tractor is shipped or sold during that time it can be wiped off with an oily rag and the appearance will be substantially that of a new tractor." While they settled on new paint technology, the question of paint color arose. "The [new] factory equipment will permit changing the color to red or any other color at any time without affecting the equipment. The question of color is not up for decision at this time. The present plan is to paint the same color in synthetic paint as is now used in color varnish."

The 4.75-inch bore and 6-inch stroke inline four-cylinder peaked at 37.3 drawbar horsepower and 51.8 horsepower off the PTO, nearly the same as the gasoline version but using much less expensive fuel.

ON MAY 21, FOWLER MCCORMICK, as foreign sales manager, joined Ed Johnston's NWC meeting called to discuss reestablishing tractor manufacture in Europe. Their prime candidate was the F-12.

"In Germany we are shut out from importing tractors," McCormick explained. "There are several quite large manufacturers there; the internal trade is good. In France, it would be a great advantage when trade picks up to have a tractor for sale, which was manufactured there. We have several million marks in Germany," he continued, "which cannot be sent out of the country. Industrial knowledge and ability are available there." He wondered about engines, however. "One thing to be considered is whether we can ship engines into the country in which we are to manufacture, at least for a time. Or can we purchase a suitable engine in that country?" McAllister leaned toward opening a factory in Germany. Fowler McCormick agreed, "But we cannot tell anything about Germany until we know what the government will do."

The overworked NWC met again to hear Sperry's update on European manufacture on January 2, 1936. Touring Germany since Christmas, he learned that *treibstaff*, a combination of gasoline and alcohol "of better quality than diesel fuel," would work well in the gas/kerosene-engined F-12s.

GROWING ACCEPTANCE OF PNEUMATIC RUBBER TIRES forced GPED to reconsider tractor speeds. Before this March 16 meeting, top gear speeds had developed haphazardly. Now IHC needed continuity across product lines when each agricultural tractor was tested at Nebraska.

1938 F-12. It's not an extremely early version of no-till farming but simply a cultivator-equipped F-12 posing for pictures in harvested corn stubble in central California. IHC manufactured more than 120,000 of these models over its six-year lifetime.

Sales brochures and service manuals from the Advertising, Sales, and Service Departments described each machine fully. Numbers had to match and speeds had to increase.

Sperry, veteran of several Nebraska tests by this time, pointed out that the actual traveling speeds recorded in the tests varied due to lugs and ground conditions. David Baker revealed that IHC's engineers (and their competitors) selected lugs literally for the day's conditions on the test track. He reminded his colleagues that it was necessary also to have adequate brakes for such speed: "Such brakes would soon be made necessary by legislation."

The first of five preproduction F-21s (the F-20X) reached Phoenix in early April. Two remained in Chicago for Deering Works to design implements for their hydraulic lift mechanism. Early reports were very favorable. One F-21 in Phoenix ran 145 hours with a three-bottom plow working 10 inches deep. With GPED and NWC both watching tractor weights, Sperry felt compelled to explain why this F-21 weighed 400 pounds more than production F-20s.

"The new 54-inch wheels required heavier-spoked rims," he said. "The extended axle brought in additional weight. The larger diameter wheels effectively reduced the wheelbase by seven inches, so the wheelbase grew six inches to accommodate front cultivators. Part of the added weight of the F-21 came from using cast iron in lieu of more expensive material. Transmission weight came from the need for a 15-mile-per-high high speed on pneumatic tires.

This very late production tractor was the closest to perfect that IHC engineering could make the model. At one point, facing its seemingly endless engine problems, Leonard Sperry proposed stopping production until engineering could find solutions.

"Management insisted on carrying over the F-20 engine into the F-21. It could produce additional horsepower without much increase in weight, and up to 25 horsepower at the belt would be possible." Sperry said the new F-21 engine weighed 958.5 pounds (with a new head, manifold and air cleaner) just two pounds more than the F-20 power plant, no penalty for nearly four more horsepower.

"If we are on the wrong track as to the type of tractor," Sperry said, "we must redesign it. There is nothing much we can do to reduce the weight of the tractor now proposed for the purposes for which it is designed." However, field tests during design phases were disheartening. A John Deere Model A gained one round on the Farmall in every 10 rounds plowed. While the F-21, the F-20 replacement, was nearly 900 pounds heavier than the Deere, Sperry argued that its higher speed, quick implement-mounting system, now referred to as Quick Attach (QA), and its improved plowing power moved it past the competition.

If they could keep too many more new ideas from infecting their designs and slowing development, Johnston and Sperry both believed 1937 would be a strong product year. They had a new Farmall and two new crawlers. By early August, Sales pressed GPED to adapt adjustable rear tread to the F-30 as well as the F-21, now referred to as the New F-20. Sales interfered again with Engineering, proposing in mid-October that, instead of just adopting

1938 F20. As IHC engineers tried ever harder to increase Farmall engine output, R.M. McCroskey experimented with using high-test gasoline in F-20s early in 1937. Ed Johnston balked, worrying that "it would be foolish to put more power in the engine than the chassis or power train could handle."

adjustable rear track-width mechanisms to the F-30, GPED should redo the big Farmall along the lines of the Intermediate New F-20, or F-22. Engineering also began work on a new tractor to replace the W-40. From the start, Sales had considered the W-40 a temporary tractor, even referring to it as the "converted 22-36 tractor."

Back in April 1935, A. W. Seacord had suggested the change from gray varnish to synthetic enamel paint would let a tractor's color last longer when stored outside. As the Experimental Department worked with sprayed paint techniques, they tested various colors in May and June. The industry already had enough green tractors between Deere, Oliver-Hart Parr, and some of their

own lines. Case and Ford seemed devoted to gray, as IHC had been, so desire for differentiation led to tests at Hinsdale farm. While Sales wanted something new, GPED wrestled with how they could make the tractors more visible, especially on roads; red continually emerged as the answer. By summer of 1936, tests showed the new red synthetic enamel held up to steady exposure to sun and elements better than the previous gray had. The EC circulated decision papers, GPED issued Change Decisions, and Manufacturing released specification change orders. On November 1, 1936, the first 1937 model tractors rolled out of Tractor Works and Farmall Works wearing "Harvester No. 50 motor red synthetic enamel paint." Wheels remained dipped in Harvester red color varnish.

IHC began production of the Increased Power F-20 in early September 1937, to keep pace with competitors in an ever-expanding horsepower race.

BE CAREFUL
F-20
McCORMICK-DEERING
FARMALL
FARMALL

IHC offered the F-20 on 40x6 or 42x12 rear tires. When Stew and Pat Thomet restored her father's tractor, Stew mounted 12.4-36s on the machine, with 6.00x16s up front.

With 220 cubic inches behind the bright red paint, and IHC's own E4A magneto providing ignition spark, Ed Johnston's engineers pulled 15.98 drawbar horsepower and 23.8 horsepower off the belt pulley in early versions of the F-20.

Modern hoses and clamps tightly contain the air to the carburetor. Increased power versions such as this developed 26.7 belt pulley horsepower, tested on distillate fuel at University of Nebraska.

Early in 1937, R. M. McCroskey experimented with high-test gasoline in F-20 tractors using high-compression pistons. Johnston objected, pointing out "that our tractors, including the F-20 Farmall, were designed with an 'engineering balance' as to power and strength throughout their various parts and it would be foolish to put more power in the engine than the chassis or power [train] could take care of." This concern would haunt IHC through the remainder of the corporation's life. Still, McCroskey stressed the importance of preparing to use the higher-performance fuel and, especially, the benefit of its improved fuel economy despite its higher costs.

"Farmers want low cost of operation," John McCaffrey argued, "which we have shown is accomplished by our tractors using low cost fuels. High test is not low cost." Yet within 18 months, GPED, which loved challenges, found it could replace standard pistons with those acquired from outside sources, meant for altitudes above 5,000 and 8,000 feet. At sea level, this provided higher compression. However, the performance pendulum swung back in February when Sperry told NWC members that nearly all tractor manufacturers who submitted machines to Nebraska ran the tests on distillate.

"With the proper compression," he explained, "this [fuel] gave in the neighborhood of four percent better horsepower than a similar tractor designed for kerosene and 15 percent better results in fuel economy." Distillate was low-cost fuel, available throughout the United States and Canada.

W. E. Payton, Service Manager at the St. Louis Branch, sent Leonard Sperry a telegraph on April 8, 1938, telling him that two prototype Allis-Chalmers Model B tractors worked fields nearby. Sperry dispatched McCroskey. He and Frank Bonnes from Domestic Tractor Sales were impressed with the semi-unitized-frame and torque-tube construction. Bonnes was perhaps more affected by the area and its small scale of farming. J. M. Strasser, assistant branch manager and his guide, told Bonnes that around St. Louis, there were more than 1,200 farms of between 5 and 40 acres, farms too small to use any tractor other than a one-row machine.

"We have, I believe, just missed this market with our 12-series tractor," Bonnes alerted the NWC. "The price of our F-12 has gone steadily upward until at the present time these small garden farmers will not make the investment required . . . . If this [Allis-Chalmers] tractor, with its main frame construction or torque-tube main frame . . . can stand up, this little tractor is certainly the greatest threat to our F-12 business today."

1938 W-30 Orchard California Special. Well-known IH collector Mike Androvich likes time capsules. These are unrestored machines that remain completely original in their equipment and their appearance. For restorers, these are great resources to see how the factory assembled them or how early operators modified them for better service.

Through a stringent weight loss program, GPED reduced F-22 weight to less than the F-20. However, under certain farming and implement use conditions, the tractor front end was too light. Johnston added 400 pounds back onto the F-22.

The updated F-14, or F-15, now had 20 horsepower compared to 18 horsepower from F-14s and 16s in the F-12. The F-14 suffered from front-end weight-and-balance problems as well. Halohan and McCaffrey, apologizing to GPED for telling them how to design tractors, proposed they lengthen the F-15's wheelbase by 6 inches to 85 inches, matching the F-22. Johnston argued this would make little difference; an additional 350 pounds might still be necessary to keep the

Stiff pressure from John McCaffrey's domestic IHC sales operation forced GPED to provide tractors for every farm and every crop. California vegetable farmers needed high-clearance models, while grape growers and fruit tree farmers wanted other special equipment.

Besides the full fenders, Orchard California Special model buyers got a tractor with a foot clutch and shorter grousers that reduced the risk of damaging fragile tree roots.

IHC's W-30 models first appeared in 1932 while its "official" orchard model, the O-12, debuted in 1934 as a contemporary of these bigger, more powerful California Special orchard models.

nose on the ground. On March 17, McCaffrey urged quick decisions so F-22 production might begin on schedule on November 1. Manufacturing had promised him 4,000 tractors available by January 15. Tooling and raw materials costs were rising, but salesmen were confident. Then, on April 13, revised manufacturing cost estimates made McCaffrey reverse himself.

"The F-20 is one of the best tractors in the Harvester line," he now proclaimed. "The F-20 is still salable if certain improvements are put into it. It will cost $2.5 million [$30 million, adjusted to 2005 figures] to put the F-22 into production and it will have to sell for $1,025 against $985 for the F-20, both on steel [$13,020 against $11,820]. This higher price will put us out of the market. And this 'dressing up' of the tractor! A farmer puts a tractor out into the field

Original equipment and even the decals hold up well in a sympathetic California environment. It's apparent from details such as this that while the tractor saw use, it was not abused.

to work. This dressing up does not mean anything. A tractor is built to do a job quickly and cheaply. Too much has been done on the appearance factor." He referred to outside industrial designer Raymond Loewy's sheet-metal treatment on the new prototypes. GPED design engineer A.W. Scarratt was justifiably incredulous.

"The F-22 is the result of unfavorable comments from Sales on the F-20 when compared to competitive machines," Scarratt exclaimed. "As a result, features were developed which were incorporated into what is now the F-22. This machine has all the things that sales have asked for: clearance under the machine, adjustable tread, new transmission to offer speeds from two-and-one-half to twenty-one miles per hour, provision for pneumatic tires, variable speed governor, good brakes, better steering wheel, water pump, foot accelerator, a standing platform, and it has the increased power everyone asked for.

"All these things were being developed," Scarratt continued, "and the effect of style consciousness crept in and so we dressed up the job. The fact of our having 'styled' this job has caused no penalty in the tractor mechanically and this sheet metal can be taken off.

"But our competitors do not have all the improvements in one machine that we have in the F-22," McCaffrey replied. Sales felt wary of being first with so many new features. "How many of these things could be put on the F-20?"

David Baker explained that while GPED might save some tooling cost, redevelopment would still be expensive. McCaffrey wasn't swayed.

The W-30 was supposed to replace the 10-20 but IHC continued its production throughout the W-30's lifetime. The W-40 replaced the aging 15-30/22-36 gear-drive models. The newer versions carried over their predecessors' spring seat and covered PTO shaft.

"The adjustable axle tractor business [is] less than 20 percent of the whole," McCaffrey observed. "We now have the F-14 which will supply the demands of part of the 20 percent. About 70 percent of the tractors in the field never adjust their axles." R. C. Archer, manager for tractor sales, jumped in: "Deere and Oliver are selling tractors because it is possible to adjust their wheels."

Then Ed Johnston arrived, joining a meeting that threatened, unexpectedly, to undermine three years' work, much of it done at McCaffrey's request. "The question of revamping the F-20," Johnston said, "is whether or not Sales can get along without the adjustable tread. If we start to remodel the F-20 it will mean we have to change the whole thing and when we are finished we will still have a tractor which will not be considered new by the trade." Johnston's clarity and insight had no impact at all. The NWC met at Tractor Works on April 27 and killed the F-22, creating instead a new three-plow tractor, the F-32.

"The term F-32 has a different meaning as to power and size than previously," Ed Kimbark reported in meeting minutes, "when it was tentatively assigned to a redesigned F-30 with an engine having horsepower of about 40." By the end of the meeting, the NWC endorsed a one-, two-, and three-plow tractor, but with new designations. The one-plow F-10, authorized in September 1937, was renamed the 1-F by GPED, to avoid mushrooming confusions with Farmall F-number designations. The two-plow F-15 with a new 22-horsepower engine

Another feature of the California Specials was their heavily shrouded intake and exhaust manifolds. These meant to protect young branches from damage as the tractor rolled past.

became the 2-F. The F-32, its name actually used for a short while on a proposed "22-36 Increased Power Farmall" (authorized back in April 1936), became the 33-horsepower F-32 three-plow Farmall. GPED renamed it the 3-F. The NWC accelerated prototype work already under way with older numbering. The F-10/1-F was due within weeks; NWC expected the 22-horsepower F-15/2-F in June 1938; and by July, the Committee wanted the F-32/3-F. Johnston announced that he would have manufacture-ready prototypes in July 1939, "this date being dependent on sales not changing their minds again as to what they want in sizes and powers," he added. The meeting minutes did not reflect the tone of his voice.

For the F-10/1-F, Johnston proposed an L-head engine and a new overhead-valve version. The F-15/2-F provided drawbar performance to match Allis-Chalmers' WC. McCaffrey, again apologizing for designing tractors, said that the F-22, at 29 gross horsepower, was not enough to call it a three-plow tractor. "A three-plow tractor," he explained, "should have at least 31 and better, 32 horsepower."

Johnston reminded him that the 33-drawbar horsepower F-32/3-F would weigh perhaps 200 pounds more than the F-20. However, few parts of the F-22 could be used on the now-proposed F-32/3-F. Charles Morrison wondered aloud, "Under the present plan, then, does the F-30 just fade out of the picture? And what about a four plow model?" Ed Johnston described to them two tractors known in GPED as the F-40 and the W-42.

IHC powered these tractors with its 3.75 x 5-inch inline four-cylinder engine with 220 cubic inches displacement. In tests at Nebraska University, the F-20s developed a maximum 26.7 horsepower on the belt pulley and 19.6 off the drawbar running on inexpensive distillate fuel.

Michigan collectors Bob Findling and George Morrison work carefully to restore the machines they own. The IH E4A magneto glistens in the late afternoon sun.

"The engine proposed [a new 5x6.125 four-cylinder with dry liners] was in two forms, one with the engine base as part of the tractor frame and one with the engine to be mounted on a more conventional chassis structure. The W-42 would be a real four-plow tractor."

"We must have a four-plow tractor," McCaffrey concluded, never one to let a new product idea pass in silence.

WHILE FARMERS DEMANDED LARGER IMPLEMENTS and Bert Benjamin's implement makers responded, Ed Johnston's tractor engineers encountered problems with the power lift mechanisms. Scarratt and Benjamin used 5-inch-diameter hydraulic cylinders (replacing previous 3-inch sizes), mostly to accommodate heavier F-30 equipment. Carl Mott designed a system that lifted left or right gangs or front or rear sets independently. A double-acting cylinder valve-and-spring mechanism raised and lowered front and rear equipment in sequence.

GPED discarded the overhead-valve engines in F-10/1-F prototypes. Sperry cited large tooling expenditures as the primary reason. NWC authorized two more prototypes in late June 1938, one using an L-head engine and the other with a redesigned F-12 power plant for comparison purposes; Sales kept pressure on Sperry to complete testing in time to launch production on March 1, 1939. They calculated that if GPED followed customary preproduction programs, including several-month lag-times between testing, prototype costs, and design changes, "our competitors will have sold 40,000 tractors before we even enter the market. Allis-Chalmers are building and selling 75 and perhaps as many as 100 Model B tractors per day. When we lose tractor business we lose everything that goes with it: plows, cultivators, disk harrows, mowers, harvesters." Then H. D. MacDonald, from sales, confused things. While pushing an accelerated testing program, he also insisted on adjustable front and rear axles to straddle two rows.

Standard-front F-20s stretched out 140 inches in overall length and stood more than 80 inches tall on pneumatic rubber. The tractor weighed 3,950 pounds at the factory.

The NWC and EC met at Hinsdale on August 2, 1938, to watch several demonstrations, including a mock-up of a 2-F tractor plowing against an Allis-Chalmers WC, both using two-bottom 14-inch Little Genius plows. Immediately afterward, the participants began to second-guess the prototype.

Sydney McAllister asked, "Are we on the ragged edge as to horsepower? We must consider having ample power for two plows and also we must consider cost!"

Ed Johnston, frustrated by company management that wanted heroic efforts in development, and miracles in meeting deadlines while juggling too many projects in the air, snapped, "If more horsepower is wanted, we must start all over. IHC's outside industrial designer [Raymond

continued on page 149

1941 O-4. These orchard-fendered models weighed 4,320 pounds, about 430 more than the standard W-4 models that IHC manufactured at the same time. In all, the company produced about 2,721 of these compact orchard tractors.

This inline four-cylinder engine displaced 152 cubic inches with cylinders measuring 3.375 inches of bore and 4.25 inches of stroke. IHC conservatively rated theses machines at 25 horsepower on the drawbar and 27.5 off the belt pulley or PTO, burning gasoline.

With equipment ranging from a large front (and small rear) light to its PTO-shaft-driven, rear-mounted side cutter bar, this tractor was meant for business then and exhibition now.

1939 W-14. Sometimes, as a collector, you get lucky. Mike Androvich found this W-14, equipped with nearly every single option that IHC and its dealers had available at the time.

IHC produced about 1,163 of the W-14 models. They sold new for $720 on steel, or $920 on all rubber.

Originally sold on steel wheels, IHC offered pneumatic rubber starting in 1940. Many farmers retrofitted earlier tractors with the easier-riding soft tires. This one sits on 11-40 rears and 5.50-16 fronts.

Today, the original specification 39.375-inch front wheel tread sits on knuckles and bolts coated with a gentle patina of rust.

Continued from page 142

Loewy], has approved the proposed styling of the tractor though it is not just what he would like to have, which, we've determined, would be too expensive."

"It would be very desirable," McCaffrey mentioned at the end of that meeting, "to announce all three Farmall tractors to the trade at the same time." The board settled on phase-out plans for F-30, F-20, and F-14 models; new Farmalls would require factory production space, and markets for older tractors would surely end after introduction in September 1939. On

The W-14s measured 50-inches wide without the overhanging rear cutter. At the factory, before options and riding on steel, these tractors weighed just 2,900 pounds.

1939 F-14. At the other end of the spectrum from unrestored original tractors is this sparklingly restored jewel. Michigan collectors George Morrison and Bob Findling labor long and hard to get their machines to look this fine.

October 3, they agreed to drop the W-30 and 10/20 tractors before May 1939, to meet demands on factory capacity.

Then, on November 9, during a conference at Hinsdale farm, another McCaffrey issue returned to haunt the NWC members. The previous March, he argued that adjustable-tread tractors made up only 20 percent of the market and were not needed on 1-F tractors. That was before the NWC saw the Allis-Chalmers Model B with adjustable-track widths from 40 inches to 52. Allis sold nearly 11,000 tractors by year-end.

Fortunately, David Baker already had an adjustable front and rear axle. McCroskey now reasoned that perhaps 90 percent of the 1-Fs would be sold with these axles. When Ed Johnston showed a prototype row-crop 1-F, with a single front wheel to accommodate two-row planters or cultivators, McCroskey feared the future:

"When purchasers had, in this size tractor, the ability to cultivate two rows," he said, "they would want it capable of using larger implements. There would be an insistent demand for more power, followed by [our providing] more strength and weight." McAllister and Morrison agreed, telling Johnston to continue experiments to "protect ourselves by perfecting a three-wheel type 1-F, but it must not be offered to the trade unless it is decided to do so later by force of circumstances."

By "circumstances" they meant the pressure of competition. During a June 13, 1938, conference on "Development of New Machines," John McCaffrey set out six guidelines for information he required before Sales would consider new development or major redesign projects. They were:

1. Competitor's weights
2. Competitor's list prices
3. Our weights required
4. Our required list price
5. Our product cost necessary to establish this list price
6. Estimated sales

Owners and operators could adjust front tread width by as much as 8 inches per side. This optional front axle added $65 to the $655 base price in 1939.

The F-14 engine was essentially the F-12 engine that normally ran at a rated speed of 1,400 rpm but in this new designation ran at 1,650 rpm. This increased horsepower output from a peak of 11.1 on the F-12 drawbar to 13.2 for the F-14.

Rear fenders added $15 to the original price of $800 with pneumatic rubber tires. With its adjustable wide front end, it is far less than gross F-14 production figures of more than 27,000 tractors would suggest.

Projects stood a better chance of Sales Department endorsement if someone else already made it at a price IHC could meet or beat. Fowler McCormick had gained influence here: After five years as head of Foreign Sales, the board named him vice president of Manufacturing in 1938. His staff determined costs that influenced IHC's retail prices.

Throughout that winter, GPED continued development work and then IHC launched a number of new tractors. On May 23, 1939, Baker signed off on production orders to start

1939 Fairway 14. International assembled just 114 of these higher-power Fairway models throughout 1938 and 1939. These compact models measured just 105.5 inches long overall and stood only 50 inches high.

As with the F-14 and W-14 models, one of few differences between this Fairway 14 engine and those in the 12-series was that IHC engineers increased the engine's operating speed from 1,400 rpm to 1,650 rpm, gaining about 3 horsepower as a result.

assembly of 1-F tractors, now called the Farmall A. The first machine rolled out on June 21, 1939. On June 6, Baker approved the new 2-F as a Farmall H and the 3-F Farmall as the M. Model H and M assembly began July 3. Implements designed for the three lines went on sale on July 20. Baker released the Farmall B on August 8 and the first one emerged on September 5.

The following year, on August 12, Baker signed off on a Narrow Tread Farmall B, providing rear tread widths adjustable from 56 inches to 84, in 4-inch increments. Manufacture began October 15. Nine days later he launched production of a High Clearance Model A, the AV. IHC manufactured the first one on January 10, 1941. For the time being, the Sales Department was satisfied.

IHC's engineers devised a $37 optional belt pulley that ran off the PTO shaft for those operators who needed a belt. It's a very long step onto the operator's platform.

Fairway 14 models originally appeared on wide steel wheels with "sod-puncher" lugs, small rounded cones that aerated the lawn as the tractor drove over it. Many operators converted to pneumatic rubber.

PREWAR INNOVATION AND PRODUCTION SURGES FORWARD

Chapter 6

1940–1944

The New Work Committee pushed along one large engineering project after another throughout 1941, releasing them almost every other month. Most involved diesel engines. During a Farm Tractor and Implement Group (FTIG) meeting February 27, Leonard Sperry, Sydney Morrison, John McCaffrey (just named vice president of worldwide sales), and soon-to-be-elected IHC President Fowler McCormick (succeeding his father, Harold) all agreed to build a sample Model H high-clearance tractor with a diesel engine.

However, while the NWC kept busy and IHC was productive, its work force labor was not happy. The day after the diesel high crop approval, some 6,500 workers struck McCormick Works. By March 3, nearly 15,000 employees sought representation by the Farm Equipment Workers organizing committee (FEW). Strikers returned to work March 23, content for the time being.

IHC produced more than 117,000 of these small farm tractors between 1939 and 1948. In 1940, the base machine sold for $515.

McCORMICK
FARMALL
A

A small group gathered in GPED in late September to discuss the Farmall B straight-axle tractor and a new Farmall E model that incorporated live hydraulic implement lift and independent PTO. Because of the B's older drop-axle technology and difficulties in updating it (necessary two years later when IHC revised its Hs and Ms), the NWC decided to go ahead with the E instead, incorporating the Model B's straight axle for this new prototype. Throughout this project, Tractor Division Experimental Engineering continued mostly doing peacetime developments even as IHC devoted more factory space to armament production.

On December 7, 1941, the Japanese air force bombed Pearl Harbor. President Franklin Roosevelt declared war the next day. World War II became an American reality and the semiannual Tractor Conference discussed shipping new tractors on steel wheels because the government curtailed rubber deliveries. IHC had a five-week supply of pneumatics remaining. Sperry and David Baker worried that steel-wheel technology no longer was adequate for more powerful

1943 McCormick-Deering Model A GPED engineers continued developing the hydraulic implement control systems, assembling test versions as they worked. This Model A prototype shows the Frame-All Touch-Control hydraulic lift and battery box.

peacetime tractors. GPED devised stronger steel wheels and recommended 5-mile-per-hour transport speed limits.

On January 2, 1942, A. W. Scarratt showed the NWC a wooden full-scale model V-8 engine developed at Fort Wayne. GPED designed the V-8 in two displacement sizes, one capable of a maximum of 460 cubic inches and a larger one up to 655. "The desirability of compactness in tractor applications," he explained, "is nearly as vital as in truck usage." Fowler McCormick wrote to A. W. Scarratt on March 7, 1942, reflecting on the proposed V-8s. "Our experience in the past has not been fortunate [using] tractor engines in trucks, or vice versa. If we set out from the start to design an engine suitable for both, we will so compromise the design to make it adaptable that it will be less desirable for either."

To Scarratt, the more painful compromise came from products developed from Sales Department wishes that discouraged innovation. GPED improved competitors' machines so

1943 Model O-4. IHC offered its smaller wide-front W-series tractors in an orchard configuration. This provided full rear fenders, under-hood air-cleaner, and exhaust. Here the 152-cubic-inch 16-plus-drawbar horsepower model prepared furrows in a California avocado orchard for irrigation.

1944 Model M. IHC's new Model M was the next tractor to get Touch-Control development. GPED got this prototype, with its experimental hydraulic lift levers, photographed on February 5, 1944.

IHC could put its label on them. McCaffrey's six rules codified this, even as he apologized repeatedly for "appearing to design the tractors." With this V-8 engine project, Engineering tried to take back engineering.

GPED shelved the Farmall B Straight Axle tractor on January 16, 1942, due to wartime need for raw materials as well as GPED's Farmall E development, which they saw as a pilot model for future tractors. The war also sidetracked QA hitch development and the QD Quick Detach system that Bert Benjamin had revised the previous August. (The straight-axle B eventually would reappear in 1947, introduced after the war as the Model C.)

Arnold E. W. Johnson offered a GPED innovation on March 25: "New machines must be designed to fit the crops and farming practice of numerous individual localities. Direct-mounted implements on basic front and rear frames may improve this. To each we can attach a variety of working tools [that] can be removed intact with the adjustments preserved, saving the operator time required for resetting them . . . . Hydraulic fingertip depth control [and] power-lift eliminates varieties of hand lever assemblies . . . . Fingertips regulate the depth of working tools. Two double-acting hydraulic cylinders lift, [or] apply pressure to the tools. Individual tools and frames detach from the tractor without the removal of a single bolt. Only one wrench is necessary."

1947 McCormick-Deering Farmall Cub. As part of the Farmall Cub launch, dealers around the country staged events to introduce the compact tractor to potential customers. Three years later, in 1950, the Louisville plant would resort to white paint to get customer attention.

1945 Model B. This photograph, made on March 6, 1945, shows how the Frame-All implement control system mounted to the tractor and how the implements attached as well. The Touch-Control–prototype hydraulic control module sits on the tractor frame next to the offset steering wheel.

1939 Model H. IHC manufactured its first Model H tractor on July 3, 1939. It quickly became the best selling Farmall. Farmall Works produced more than 10,000 in 1939 alone.

IHC had developed power take-off in 1918. Almost without interruption the company offered it as an option since 1921. Rear PTO appeared as an option on the new Model H.

This was the row-crop companion to the W-4 and O-4 series. It used the same 152-cubic-inch inline four-cylinder engine. Power output was identical.

Engineering had christened this system the "Frame-All," but recognized that this name would be replaced by one less likely to cause confusion with the Farmall tractors. Unlike the Ford-Ferguson three-point rear hitch with one cylinder and one control lever operating implements at the rear of the tractor only, Frame-Alls used two double-acting cylinders and two levers, allowing adjustments on one side of the tractor or the other, front or rear. It could accomplish delayed lift and drop of front and rear implements. Two days after the Frame-All Model H introduction, Fowler McCormick clamped tight security on it and the QA system with hydraulic controls. He theorized that "After the war, farmers would have used up equipment and would need something new. Sales would want something new to sell." Scarratt suggested planning immediately to replace A and B models with this "E" straight-axle tractor (the Model C). "Adaptations of the 'QA-Frame-All' and hydraulic fingertip lift control should be confined to Farmall E," he argued, "because the basic principles of this scheme of tools and attachments lends itself readily to this tractor which is more similar to the H and M in general outline than either the A or B tractors."

War production consumed most of IHC's factory capacity. But once tooling was set up, GPED experimental engineers felt little pressure to develop new machines every 90 days. They had the time to perfect the Frame-All and the "E." By late fall 1942, GPED had completed Scarratt's first Farmall E, and they started assembly of a "Reverse Direction Super Farmall E."

continued on page 169

ORMICK-DEERING
RMALL

This was the right tractor at the right time. It would share a wheelbase with the larger Model M so that Bert Benjamin's implements function interchangeably. This meant that farmers didn't have to purchase two sets of tools if they owned both an M and an H.

Norm Walton has a reputation throughout Michigan's thumb for his collection of H and M tractors. This H, #984, is one of the earliest known.

1939 Model M with Elwood Front Wheel Drive. Elwood Manufacturing began producing four-wheel-drive kits in the mid-1950s. Farmers needed increasingly efficient methods and equipment to get engine power onto the ground. More power to the rear wheels alone sometimes resulted in wheel spin.

Delco-Remy electric starting and lighting was another option for these modern machines in the 1950s.

1941 Model AV. The Model A vegetable tractor provided owners and operators an extra 5 inches of ground clearance over the standard A. These little tractors measured only 115 inches long and 69.25 inches high at the steering wheel.

Elwood offered these conversion kits for Farmall M models as well as for other manufacturers in the late 1950s and 1960s. By the early 1960s, these could cost as much as $2,200, a sizable investment for a tractor costing not much more.

Continued from page 163

T. B. Hale, in regional tractor and implement sales, addressed a dealer's meeting in Dallas on March 23, 1943. It had taken IHC five years since April 1938, when Frank Bonnes visited the St. Louis area, to form a product plan based on smaller tractors for mid-America's smaller farms. David Baker joined Sales and Manufacturing managers to host meetings from New York City to Los Angeles in late March 1943. The executives questioned 124 regional branch managers and dealers about IHC's current equipment and its uses, the competition and its advantages, field service and factory changes, mistakes of the past, and rumors about the future.

"The consensus was that many wartime developments might well be carried over into the design, production and uses in commercial industrial power industry after the war," Baker told Ed Johnston and Fowler McCormick. "Dealers leaned toward expanding the line upward in power and they expressed need for diesel power units as high as 200 horsepower."

Then Sperry and others distributed an eight-page "Survey of Potential Demand for Farmall 'X' Tractors (smaller than the Farmall A)" on July 9, 1943. They acknowledged the Farmall A was appropriate for farms from 40 to 70 acres. However, the 1940 U.S. Census revealed that of 5.7 million farms reporting crop acreage, 3.3 million (or 58 percent), were smaller than 40 crop acres. Of those, 2.2 million had an annual gross farm income of $400 or more (about $4,000 adjusted to 2005 dollars).

Sperry wrote that "In the 15 years of Farmall tractor type selling, IHC has sold 733,000 Farmalls." The potential, therefore, was "to reach the untouched market demanding smaller equipment." The Farmall "X" would "do the work of two or three horses or mules, be a four-wheel tractor, row-crop type, with eight horsepower on drawbar. It would be designed 'CultiVision-style' [engine offset to the operator for better crop visibility] for QA Quick Attach machines for truck garden and field work whose retail price is not to exceed $400." They proposed a complete line of implements, tools, and attachments. To keep manufacturing costs at around $213 (approximately $2,100 in 2005), necessary to meet a $400 list price, Baker and Sperry, while preferring a four-cylinder engine for torque and running smoothness, proposed a new two-cylinder, parallel, upright engine for the "X."

continued on page 176

Both the A and AV used IHC's C-113 inline four-cylinder engine with 3-inch bore and 4-inch stroke. These developed 17 horsepower on the drawbar and 19 off the pulley or PTO.

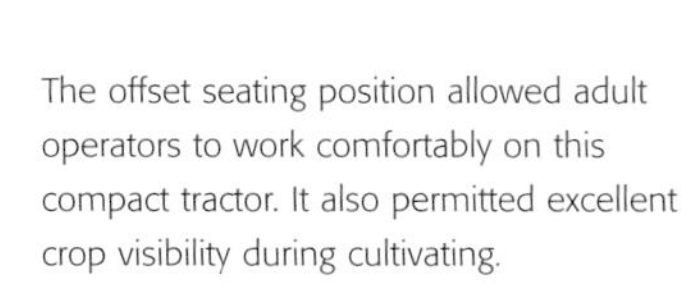

The offset seating position allowed adult operators to work comfortably on this compact tractor. It also permitted excellent crop visibility during cultivating.

1941 Model H. IHC developed the Model H at the same time and in the same testing venues as it did the M. This H, with 25.5 drawbar horsepower, made easy work of plowing in dry hard Indiana soil.

The Model H sold new for $855 on pneumatic rubber tires in 1940. With five forward speeds, it easily pulled two bottoms in second or third gear, at 3 or 4.25 miles per hour. Transport speed reached 15.625 miles per hour.

1941 Model A. The seating and steering offset is apparent in this front view. Adjustable front and rear tread width made this compact tractor versatile and valuable to small-farm operators.

The adjustable front end allowed a range of tread width from 44 up to 64 inches. The tractor weighed just 1,870 pounds.

McCORMICK
FARMALL
A

With four forward speeds available, operators could transport as fast as 9.625 miles per hour. Rear tread adjusted from 40 inches to 68.

Continued from page 170

Farm Tractor and Implement Group (FTIG) released specifications for the Model E/C. It ran the basic Model B engine at 1,650 rpm, with pump-circulation engine cooling. The E/C-chassis used a straight axle with 36-inch rear tires, 15-inch fronts, 21-inch ground clearance, and adjustable tread from 48 to 84 inches on an 86-inch wheelbase. The transmission provided five-forward and five-reverse speeds. The E/C offered continuous running PTO and hydraulic lift pump with Touch Control; it added the simplified and improved QA implement mount system, and had "styled and non-vibrating sheet metal enclosures, over the engine and rear fenders with cutouts for implement movement."

On September 1, 1943, while development continued on V-8 diesel and gas engines for trucks and tractors, McCormick's engineers attended to a full range of tractor products for agriculture, industry, and construction. Fowler saw stressed managers pulled taut by the workload and variety of projects they had to manage. Like his grandfather, he recognized people's strengths and weaknesses. IHC had strengths, too, but its tendency to force division general managers to do too much, he feared, could harm the company. In meetings he had heard lapses and errors. With millions of dollars at risk in development and tooling costs, he knew mistakes could devastate a corporation's health. McCormick conceived a plan to reorganize IHC where areas of expertise and interest influenced a manager's job selection. He separated Farm Tractors, Farm Implements, Industrial Power, Motor Truck Division, and the general line, including

refrigeration and other products, from the tightly centralized rule his grandfather had established. He wanted first to reduce executive workload while giving autonomy to division vice presidents. Second, he intended to diversify the EC, which was top-heavy with eight former sales managers as members but only one engineer and one manufacturer.

Running on gasoline, the little A developed 17.4 horsepower at the drawbar and 19.1 on the pulley or PTO. This was peak production year with Chicago Tractor Works turning out 22,950 during 1941.

At the new Tractor Division, testing the Intermediate H and M tractors with Frame-All hydraulic controls revealed that the larger twin hydraulic cylinders were inadequate for middle busters or four-row cultivators. Operators found control levers poorly placed and discovered they needed to hold onto them until the lift or drop was complete. As development wore on, E. F. Schneider from Sales grew impatient with discussions of further testing. "In recent years competitors have put new machines into production after a test with only one experimental

1941 Model AV. IHC data lists the higher clearance Model A tractors as 9 inches longer overall and nearly 400 pounds heavier than the standard clearance Model A tractors. The company produced 3,603 of the AV models.

1942 Model H. The war claimed all industrial copper and rubber. This quickly returned tractors to steel wheels and crank starting.

machine. We should release our new machines, whenever possible, without the usual preproduction lots, in order to be in the lead rather than to follow others." Industrywide, manufacturers began to rely on customers to complete their final development programs.

JUST BEFORE CHRISTMAS 1944, on December 20, in the Motor Truck Sales Room on the southwest side of Chicago, FTIG showed off the Farmall "X" prototypes to nearly four-dozen managers from Engineering, Sales, Manufacturing, Service, and Executive committees. Scarratt played master-of-ceremonies, introducing the Farmall X. Nothing on the new model came from Farmall A or B tractors because those parts could not meet weight, size, and cost specifications for the X, which they sometimes referred to as the "Baby Farmall."

McCaffrey's Frame-All program was to add 25 A and B tractors to the tests. It nearly brought manufacturing to a halt on January 2, 1944. A and B assembly-line production was

Model H tractor production remained high during the war years. IHC produced 34,987 in 1942, 21,375 in 1943, and 37,265 in 1944.

Inset: IHC eliminated fifth gear with steel wheel models. Top speed in fourth gear offered 5.375 miles per hour instead of 7 miles per hour with pneumatics.

under way; Manufacturing had orders for 5,000 of them through 1944. Interrupting this to hand-assemble 12 tractors was impossible. David Baker understood the need to complete Frame-All hydraulic control system tests but felt "manufacturing should cooperate fully to become familiar with this important new development." McCaffrey, just elected to the board of directors as an IHC second vice president, had aimed at July 1, 1944, as the production approval date, based on Frame-All models testing nonstop through May and June. Now, it appeared he would have to reset this to July 1, 1945, unless major effort moved the program.

Manufacturing was adamant. It would not interrupt two assembly lines for a single A high crop and four BNs. Fowler McCormick agreed to slip introduction back to 1945, but he reminded manufacturing vice president H. K. Kicherer that Canton Works had a variety of 25 Frame-All implements in A- and B-tractor sizes that needed thorough testing as well.

Early 1942 H production still meant a few of the tractors escaped with electric start. The base tractor sold for $695 on steel.

1943 M-LPG. Here's an unrestored original awaiting a long winter and a visit to the restoration shed. The Model M first appeared in mid-1939 and it was IHC's powerhouse.

Weaver & Lingg Implements in Sturgis, Michigan, performed more than 140 liquified propane gas (LPG) conversions in the 1950s. Howard Weaver was an aggressive advocate of the inexpensive yet powerful fuel.

One drawback of the propane gas system was the intrusion of its huge pressure gas tank. The advantage was that the farmer could use one fuel to run the tractor, heat the home and cook the meals.

1943 Model H road roller conversion. Keith and Cheri Feldman and their parents have steadily been building a collection of interesting Farmalls. This road roller-conversion is one of their more unusual pieces.

McCormick then ordered Kicherer to "make an immediate, intensive study of the hydraulic system and gain the necessary knowledge of, and experience with, this unit preparatory to its production."

THE FARMALL E/C PROGRAM, launched in September 1939, was in jeopardy in early January 1944. FTIG had converted a standard "drop-axle" Farmall B into a straight-axle tractor. This required costly revisions when testing revealed strength problems and implement mounting constraints. As a result, in late September 1941, FTIG had created the new straight-axle Farmall E/C prototype that it demonstrated in mid-August 1943. Now, getting cold feet, the EC ordered FTIG to greatly reduce cost and weight, and eliminate "certain features it felt [it] could not afford to incorporate in a tractor of this size and capacity."

"The Farmall E," Baker reminded FTIG members, "was the pilot model for a complete line, because it would be restyled and incorporate features which it was felt were essential for an up-to-date future line of tractors. Such a tractor," he proposed, "would become the fore-runner of a new line. The M size has sufficient horsepower to satisfy power requirements for a complete

line of attaching tools which are considered in new implement developments, including [mounted] harvester threshers." The final coffin nail, however, came from Archer in Sales.

"Because the Farmall E was larger than the A and B, but smaller than the H and M, [it] would render both larger and smaller tractors obsolete prematurely, thus interfering with the sale of these tractors pending new developments." This was crystal-ball gazing, similar to what Sales had done worrying about McCormick-Deering 10-20 models after introducing the Farmall.

The war strained away time and energy; existing programs soldiered on, sometimes gaining unanticipated results. Just as during World War I, while the male population fought, wives and daughters ran the farm to feed and clothe the world. IHC's branches organized tractor operation and repair schools for women. The EC told Chicago it needed to adapt current production hydraulic implement lifts to existing F-20 and F-30 tractors. This was not easy. FTIG assembled

The rear steel wheels measured 36 inches in diameter and 40 inches wide on each side. The overall width stretched 93.5 inches.

The front steering gear replaced two pneumatic rubber tires with two steel rollers. Each was 20 inches tall and 21.5 inches wide.

Keith Feldman dug into the history of this roller. He learned that Jacobs Farm Equipment, Ltd., of Essex, Ontario, produced just five of these rollers on Farmall H tractors.

H model tractors manufactured during wartime eliminated the top speed transport gear. IHC also reconfigured fourth gear, slowing it from 7 miles per hour down to 5.625 miles per hour for tractors on steel wheels.

one and found it required 42 separate new pieces, including fabricating an angle steel frame. The chance for leaks was great and GPED did not encourage the idea. However, because it would aid farmers at home, primarily women, GPED agreed to make it work cleanly and quickly.

FTIG had Frame-All A and B prototypes in fields by May 1944, but the EC decided instead to lease or loan the tractors "for test purposes only" and not sell them outright, fearing another farmer falling in love with a unique prototype as had happened with one of Ed Johnston's International 8-16 four-wheel-drive prototypes two decades earlier. IHC preferred leasing because it placed a financial burden on the farmer, who was more likely to use the tractor fully and report honestly any complaints and failures if it directly affected usual farm finances.

In IHC's plants, workers busy making half-tracks and torpedoes for the war effort found it difficult to remember farming as usual. Throughout Europe and the Pacific, soldiers and sailors used IHC products and those from other U.S. manufacturers in the fight for the return to living as usual. On June 6, 1944, (D-Day) tens of thousands of Allied troops landed on the northern coast of France. Within two weeks, Allied soldiers captured more than 30,000 Germans occupying areas near the coast. The war was not over but progress was measurable.

ON JULY 11, TESTS CONFIRMED success of the Farmall A single-cylinder hydraulic touch-control, and B, E/C, H, and M double-cylinder systems. Questions of manufacturing cost and retail sale price rose against manufacturing's production schedule. The start date slipped back from

The prototype hydraulic control pod was not a particularly handsome casting. Its function, however, was most appealing: It could raise or lower front or rear or left- or right-side mounted implements and return them to a preset position time after time.

July 1, 1945, to September 1, after FTIG learned they should not use, advertise, or list the term "Touch Control" relating to its hydraulic system, until IHC had it trademark-protected.

Archer from Sales met with McCroskey and Sperry from Engineering to schedule tests for both Model E/C and the X tractors. Complete Farmall A Frame-All tractors weighed 1,856 pounds; the new C weighed 2,150 and Xs were only 1,058 pounds. After calculating sales and cost benefits of the Frame-All tractors compared to standards, McCroskey reported that creating the Frame-All used new parts costing $47.01 (about $423 adjusted to inflation for 2005), while removing others worth $44.13 (roughly $396). This yielded a net increase of $3.12 (approximately $28 in 2005) in production costs. While this represented a small price, at this point, the EC envisioned a production run between 70,000 and 100,000 A, C, and X tractors.

On Thursday, September 14, FTIG engineers staged a massive show-and-tell for EC members, division general managers, and others in Tractor and Implement Divisions. Starting at 8:30 a.m., FTIG demonstrated what they believed were "sign-off" versions of Frame-All-equipped

hydraulic touch control–operated Farmall A, B, C, H, and M tractors. There were no failures, no miscues, and no disappointments.

As Implement Group engineers parked the last demonstrator in the Hinsdale farm shed at 4:30 p.m., company cars and buses loaded up and rolled out of the yard. IHC management knew it had seen its future that day. IHC would have innovations, techniques, products, and tools to sell farmers who came home from the war.

One of the crown jewels of Wisconsin implement dealer Arden Baseman's Farmall collection is this Model A prototype. While rumors persist that another one has surfaced, this is the only Frame-All known to exist.

1945 Model M. It was big, powerful, and handsome. IHC hired outside industrial designer Raymond Loewy to "style" its new letter series models that first appeared in 1939.

Loewy's prototype designs displayed full engine covers trimmed with slender chrome strips. The Executive Committee vetoed the chrome because of costs, and engineers eliminated the covers to improve engine cooling as seen on this and all subsequent production Ms.

As Focus Changes, Vision for the Future Blurs

1953 Super B-MD. While IHC produced just about 5,200 of these in the United States, records estimate the British M-diesel production at more than 900. Other than country of origin, however, the machines are very similar.

Chapter 7

1945–1954

World War II engaged Fowler McCormick's imagination. Through the diversity of wartime products IHC manufactured, Fowler came to believe his corporation no longer needed to be strictly a farm-equipment and truck maker. With nearly 100,000 workers in 17 plants, sales increased more than $100 million each year (about $1 billion adjusted to 2005 values) during the war. Fowler had money to expand product lines and to buy new factories to build them.

The Tractor Division staked its share of development money on a compact machine meant for the small two- or three-horse farm. Here the Naming Committee broke form. The letter-series Farmalls beginning in 1939 with the Model M reached Model Es in the early 1940s (with jumps to H and A). In various meetings, everyone referred to this new prototype as the Farmall X or the F. Now, in September 1945, public-relations man Art Seyfarth, patent attorney Paul Pippel, engineer Leonard Sperry, and the five other committee members named it the "Cub." IHC's Sales Department aimed 45 percent of total Cub production at the east and southeast.

MD
FARMALL

1945 Model BN. Chicago Tractor Works manufactured the first BN, or Model B Narrow (single) front-wheel tractor, on October 31, 1940. IHC conceived of it as a two-row version of its Model A.

Production began in Louisville, Kentucky, late in 1947. Nearly 135,000 rolled out over the next four years. Half the purchasers were first-time tractor buyers replacing horses or mules on farms where the Cub was the only tractor. In 1945, 30 percent of the United States, about 1.6 million farms, still used only draft animals. Cotton, tobacco, poultry, and vegetable-truck farmers favored the Cubs, as did people who farmed part-time or maintained large gardens. The largest proportion was farms of 10 to 19 acres.

At the other end of the size spectrum, Fowler's perceptive but forceful friend, John McCaffrey, IHC's vice president, found a growth industry in construction. To reward his work and ideas, IHC's board elected McCaffrey president and chief operating officer in 1946.

Competition seemed futile. Caterpillar sold $230 million (roughly $2.3 billion in 2005 dollars) in equipment in 1945, compared to IHC's $35 million ($350 million). Cat had a postwar advantage from G.I. heavy-equipment operators who learned Cat's machines and told their peacetime bosses what to order. To McCaffrey, Cat made a big target. IHC's attack came through a $30 million ($300 million) investment that included acquiring a former government manufacturing plant in Melrose Park, Illinois. In 1947, IHC's Industrial Power Department introduced the TD-24. It weighed 36,000 pounds and developed 148 drawbar horsepower. Leonard Sperry conceived it and David Baker designed the giant, providing buyers with 10 horsepower more than Caterpillar's D-8 and more sophistication than Cat's track clutch/brake system. The Sales Department, which had hurried development, was pleased.

The BN provided a narrower rear tread width than the standard B offered. It could be as slim as 56 inches. This catered to vegetable growers and truck farm operators.

1945 Model B. This was a typical Model B configuration with its cambered front wheels creating a small row-crop tractor. The B weighed 1,830 pounds, about half of the Model H at 3,725 pounds.

McCaffrey had sold trucks in Ohio before moving to IHC's Chicago headquarters and this background didn't prepare him for agricultural equipment. Throughout his career, he had little understanding of farm equipment markets. But he had a feel for construction. As with trucks, bigger was better and more powerful was more useful.

McCormick wasn't done expanding, however. IHC's line of 1907 cream separators led to coolers when 1930s laws required farmers to refrigerate fresh milk within an hour of milking. From wartime field hospital blood-coolers, McCormick envisioned a full line of refrigeration

These compact machines could spread out rear tread width to as much as 92 inches. IHC manufactured a total of 75,241 of the Model B tractors.

equipment. In 1946 he bought a plant in Evansville, Indiana, and within a year IHC turned out 200 freezers a day, then adding humidifiers and air conditioners. By the early 1950s, IHC led in freezer sales, and refrigeration sales more than doubled from 1950 through 1953.

Alfred P. Sloan, General Motors' brilliant chairman, had outlined his mission during a 1921 interview to provide "a car for every purse and pocketbook." He inspired generations of businessmen. During Cyrus McCormick Jr.'s time, IHC produced tractors for every farm and function. In the mid-1940s, McCormick paid more attention to GM and saw a separation of divisions that made sense to him as president of the "general motors" of farm equipment. Fowler had strengthened Farm Tractor, Industrial Equipment (encompassing portable power units, industrial and construction machinery), Motor Truck, and Refrigeration Divisions, as well as a Steel Division, and Fiber and Twine (which supplied his harvesters). He gave each its own experimental and research departments, sales organizations, personnel, and administrative departments.

1948 Super AV. The Super A appeared in 1947 and remained in production into 1954. These models offered an electric starter and lights, and they introduced the production version of the Touch-Control Hydraulic system.

While this provided McCormick's divisional autonomy, the duplication of so many functions and executives eventually proved extraordinarily costly.

Sloan's interpretation of decentralization left one central leader to make course corrections among the scattered divisions. Chairman McCormick, by several accounts, created a different system. Sloan was a director who trusted his vice presidents and managers to make everything but the most critical decisions or long-range plans. For these he was always and immediately available. McCormick's variation put more trust in his subordinates because the chairman often was unreachable and far away.

All the McCormicks traveled. Cyrus Sr. sold harvesters; Cyrus Jr. set up factories in Europe. Harold inspected plants worldwide. Fowler, as head of Foreign Sales, visited Europe and Russia. But not all his travels were business. Overworked during war years, he continued afterward to expand IHC's vision. To manage stress, he spent months in Switzerland with Dr. Carl Jung, the psychologist and psychiatrist who was his mother's favorite. Fowler developed pneumonia in

The Super A accepted IHC's new quick-attachment implement system. A universal-frame mounting system also appeared on the new Super As.

late 1947, and for reasons never explained, he kept this secret from his executives and directors. Even after nearly dying, he let only McCaffrey and a few others know that, in relocating to Phoenix, he was following his doctor's advice. It was something that was "not discussed." As a still-active board chairman, he took work with him—boxes of it. He communicated with McCaffrey by phone or mail. He looked into a dedicated telephone line to McCaffrey, an early attempt at telecommuting. But McCaffrey felt the $1,020 monthly charge (about $10,000 in 2005 dollars) was too extravagant. Without instantaneous communications from McCormick, McCaffrey followed his own inclinations more easily.

Critical viewers will note that the word "Super" is missing from the nose decals. While the serial number 255558 confirms it as a Super AV, a mix-up at the restoration shop set the wrong decal in place.

The Touch-Control hydraulic system consumed the limited additional horsepower that Super A's had to offer over the Model A. But few operators complained once they got accustomed to the accurate depth and lift control the system provided.

McCormick's extended absence confused his directors. They concluded he didn't care about IHC. When projects drifted off course, few people alerted him, or knew that they should. McCaffrey's job grew without McCormick there to handle future planning and budgets. Problems required quick decisions. Complications arose in cleaning and retooling Louisville Works from wartime production. Extra labor and facilities costs in 1946 and 1947 tapped budgets of raw materials for tractor manufacture. Tooling had to be stored, delaying Cub production and, later, the Farmall C. Labor costs remained but there was no offsetting income from tractor

sales. (IHC had owned 39 percent of the farm tractor market in 1940. Although its factories worked to capacity, the overall market grew, and its share slipped to 31 percent by 1949.)

A report to Farm Tractor Division on June 12, 1952, spelled out the damage clearly:

> "From 1946 up to the end of 1949, Louisville Works had a cumulative operating loss of $21,594,000 (about $195 million, adjusted to inflation values for 2005). To the end of 1952, the average investment at Louisville during most of this time has been approximately $50,000,000 (roughly $450 million). As pointed out, we have not only had no return on this investment, but have . . . very high fixed costs, depreciation alone being in excess of $3,000,000 per year (approximately $27 million in 2005)."

The figures were grim; reality was getting worse. Delays in tooling up Cub and Model C production threw off outside parts suppliers. Searching for revenues, they bid other projects

1948 W-6. With his two drive wheels on the high side, David Bradford and his son Ash, pulled three 14-inch plows across a portion of central Indiana bean field. The W-6's 248-cubic-inch four developed 32.8 horsepower on the drawbar, sufficient even for dry soil.

1948 O-6. Raymond Loewy's styled orchard tractors have been described as "Buck Rogers at 4 miles per hour." But that would only be third gear. In transport gear, this tractor could reach 14!

If any tractors can be called beautiful, they must be the full sheet-metal orchard models of the early 1950s. Form followed function, and protecting tree branches led to curved surfaces and sealed seams.

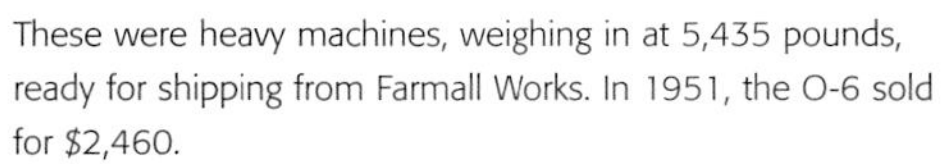

These were heavy machines, weighing in at 5,435 pounds, ready for shipping from Farmall Works. In 1951, the O-6 sold for $2,460.

The full shroud over the steering wheel protected the tractor operator as much as the trees. Orchard models dropped the operator's seat down several inches as well, to more easily clear low-hanging branches and fruit.

IHC engineers rerouted the exhaust and hung headlights as low as possible. Everything was done to protect the crop, but the result was a rare and beautiful machine. IHC manufactured just 1,962 O-6 models.

that started up more predictably. When IHC was ready to assemble the Louisville tractors, suppliers often couldn't comply. IHC's huge labor force, supported by union contracts that limited the corporation's flexibility, was not always available when parts arrived. If certain portions of assembly slowed or stopped while waiting for parts, Louisville Works managers could ask union members to work another job. The contract gave workers the right to refuse. They could go home. Then other portions of assembly backed up, waiting for workers.

1949 Model MD. John Wagner's beautifully restored diesel M graces the book cover and these inside pages. IHC built the first one of these on January 13, 1941, and continued to produce them until late March 1952.

1949 Farmall Cub on Stilts. The Tractor Stilts Company of Omaha, Nebraska, produced its first ultra-high-clearance conversion in 1948. Soon after, it began to manufacture kits for nearly every tractor make and model. Farmers used these conversions frequently for detassling corn.

One other labor problem confronted the Tractor Division, at this point run by Ted Hale, another sales vice president without manufacturing experience whom IHC's board had named a division manager. Between 1950 and 1954, Farmall and Louisville plants suffered a high, average-annual employee-turnover rate due to resignations or layoffs from production slow-downs or assembly-line changeovers. From the beginning of each year to the next, two out of three employees changed.

Fowler McCormick blamed McCaffrey for the company's failings, even among unions against whom Fowler had exercised a traditional McCormick family animosity. But other faults did lie with McCaffrey, including his almost insatiable hunger for power. In early 1951, McCormick, concerned for the future, tried to boot out his former friend. He hoped to force a

IHC's D-248 inline four cylinder used 3.875-inch bore and 5.25-inch stroke. At 1,450 rpm, the engine developed 27.5 horsepower on the drawbar and a peak of 38.2 PTO horsepower.

showdown: McCaffrey or McCormick. But the board, especially the older outside directors, resented his intrusion. They chose McCaffrey, granting him Fowler's chief executive officer title as well. Fowler was forced into an inactive Executive Council and board position. That July, the board promoted Brooks McCormick. He had moved from Melrose Park manufacturing to Truck Sales in Kansas City and then to district manager for General Sales in Dallas in 1950. Before coming to world headquarters, his job was joint managing director of IHC's British operations at Doncaster.

As the McCormicks journeyed through the world in an almost hereditary repetition of styles, so did they put distance between themselves and daily corporate responsibilities. There was, almost from the first, from Cyrus Sr. a kind of noblesse oblige, a sort of behavior that gave the message to their directors that "we have others to do this work with us. We trust their abilities as much as our own." With Cyrus Sr., it was his younger brothers Leander and William who ran things; with Cyrus Jr. it was Funk and then Legge. For Fowler it would be John McCaffrey. It was as if, even though machines and buildings were named after them, the McCormicks adhered to a genetic family blueprint that told them to trust others to manage some portions of the business in the McCormicks' best interest. They behaved as though programmed to give others responsibilities, to bring in other perspectives, to seek diversity in management, often letting others take the credit, even when the McCormicks believed they could do it better than

Electric lighting was optional on the $3,145 base tractor. The five-forward gear transmission provided a transport speed of 16.375 miles per hour.

While the Cub on Stilts provided nearly 60 inches ground clearance, the 1956 International Cub High-Clearance wide tread at right offered not quite half of that but with 78-inch tread width.

1950 Model C Demonstrator. IHC introduced the Model C in 1948. Yet a corporate promotion in 1950 launched a series of white demonstrators to explain to farmers that this C was something special. Touch-Control hydraulics first appeared on the C- and Super A-series tractors.

Inset
These tractors used the same C-113 engine that IHC fitted into A- and B-series tractors. But running the engine at 1,650 rpm instead of 1,400, and increasing compression to 6.1:1 from 5.33:1 gave the C a few extra horsepower.

those they'd hand-picked. This "shared responsibility" style of directorship built the company but ultimately hastened its end.

TOM HALES' TRACTOR DIVISION'S highest labor turnover was at Louisville, 72.4 percent annually. Production and labor conditions there led McCaffrey and the EC in 1953 to consider moving the entire production line of the new Farmall 300 to Kentucky from Farmall Works before regular production began to stabilize the workforce. A study indicated that such a move would save IHC $400,000 a year (about $3.2 million in 2005 dollars) in tractor shipping costs, but the relocation itself would cost $5 million (approximately $40 million).

McCaffrey's enthusiasm for construction equipment fueled expansion and improvements in IHC's crawler line. Impetus for transmission development came from the field. "There has been an insistent demand from the field for a hydraulic drive for the TD-24. In response, the Engineering Department has developed,

in cooperation with the Allison Division of General Motors, a combination torque converter and power shift transmission which can be introduced into the basic TD-24." Throughout late 1947, the big crawler's insufficient development time began to show. Gears overheated, failed, and in some instances, shattered inside the cases. A torque converter reduced the shock from the engine to the drivetrain. Even more important, when an operator stalled the tracks, they could keep the engine running and the hydraulic controls functioning. The first torque-converter prototype went to the U.S. Navy in April 1953 for testing. However, increasing financial constraints meant the Engineering Department built fewer prototypes and had shorter testing

IHC assembled these tractors at its new Louisville, Kentucky, plant. For three months in 1950, the factory manufactured tractors in white. Dealers could order them, complete with cardboard placards that showed off every new feature.

McCORMICK
FARMALL

1950 Model C Demonstrator and regular production version. While IHC uprated the engines slightly, it greatly increased the frame strength and overall weight of the Model C over the Model B it replaced. Both demonstrator and production Cs weighed 2,780 pounds dry while the A was just 1,870.

1950 Farmall Cub Demonstrator. IHC manufactured its Cubs at the Louisville plant. During the same 1950 promotion, Cub demonstrator models appeared all in white.

periods than necessary on new projects. Rather than slowing new product approval, John McCaffrey encouraged more construction equipment into life.

Gross sales from Tractor Division for 1953 reached $257.6 million ($2.19 billion adjusted for inflation), but that dropped by $100 million in 1954 (about $850 million) to $156.6 million ($1.33 billion in 2005 dollars). Shipments to dealers decreased by almost half, from 14,601 in 1953 to 7,952 in 1954, taking division net income down steeply from $18.9 million ($160.7 million in 2005) to $5.4 million ($45.9 million). Truck sales beat farm equipment for the first time in 1954, and would exceed it by half again in 1956. Fowler's reorganization made each division

responsible for itself; money needed to prove products before manufacture dried up. Some of this budget tightening resulted from price cuts on Farmall Hs and Louisville tractors because IHC began slicing inventories before releasing the hundred series.

In mid-1954, as introduction of the new Farmalls approached, IHC began to feel the cash flow slow down. The company responded by shipping tractors rapidly and randomly. In August, McCaffrey received letters from several branches complaining that "a sizeable quantity of tractors [had] been shipped to territories for which they were not suited." He learned from General Sales that "errors had been made in this distribution but . . . many of the tractors originally shipped

From the engine flywheel housing back, nothing is standard M equipment. This prototype boasts larger disc brakes, longer rear axles, and larger axle diameter as well as shorter housings. Engineering set the hydraulic pump inside the rear casing, one more upgrade that would not appear until 1963.

1950 M-8 Prototype. The "M-8" designation here stands for manual eight-gear transmission. This is a prototype of the dual-range four-speed system.

had been sold and [they were] confident that the remaining units would be moved in due time." Thirty years before this, as inventory of soon-to-be obsolete tractors swelled, Alex Legge had the excuse of Henry Ford's price war to move out Titans and Moguls at any price before the McCormick-Deering 10-20 and 15-30 Gear Drives arrived. Legge also had a much more solid financial footing with which to survive the losses.

By October 1, 1954, McCaffrey recognized that IHC's sales projections of $1 billion for the year were too optimistic. Peter V. Moulder, executive vice president for Tractor, and Implements Divisions (and former Truck sales manager), summed up the bad news.

"Results were very poor in spite of drastic economy measures taken, particularly at Louisville Works . . . . As you know, this [was because] estimated production for 1955 represents

only 31 percent of the tractor capacity of that works." Unfortunately, even though Louisville could produce more, there were no customers. In 1953, 42 percent of IHC's farm tractor sales were in Farmall A, C, and Cub ranges of 9- to 24-horsepower tractors; 30 percent went to Super M sales, the 40-horsepower-and-up-class tractor. The remaining 28 percent was split between H (25- to 29-horsepower) and Super H (30- to 34-horsepower) models. The new Farmall 300, replacing the Super H, would move a tractor into the previously vacant 35- to 39-horsepower bracket.

A Sales memo decreed that this move "opens up for consideration increased horsepower for the 200 model which replaces the C and will probably be in the 25- to 29-horsepower category. By increasing the horsepower even further it could be moved up into the vacated 30- to

1950 Super A with Auger. With 19.1 horsepower off the PTO, this Super A was the perfect candidate to run the long auger. These 2,385-pound $1,150 tractors were very popular. IHC manufactured about 94,000 of them between 1948 and 1954.

1951 Farmall Cub. Conceived and developed as the Farmall X or Baby Farmall, IHC introduced the Cub in 1947. The sales division had concluded that southeastern cotton and tobacco farmers could use a machine that was smaller than the Farmall A.

34-horsepower category where there is a big market." Nature was not alone in abhorring a vacuum, and so the proliferation of IHC models was self-perpetuating.

This system should have plugged every hole. Yet the broadening product lineup missed the mark. Industrial growth during and after World War II pulled 1.5 million families off farms. Many who remained bought their neighbors' land and needed bigger equipment to work holdings that encompassed a half section or more. By 1954, some 130,000 farmers or ranchers worked 1,000 acres or more. While fewer than 3 percent of the farms, their owners bought 9 percent of the tractors in the United States.

Taking cues from the auto industry, IHC, like Ford and Deere & Company, introduced a new, improved tractor model every other year. As engine developments increased tractor

More than a garden tractor, IHC conceived the compact Cub as a real working machine for small acreage farmers. Of course, this philosophy made it into the perfect tractor for large-acreage estates with lush gardens.

performance, some horsepower categories filled and others emptied. In several instances, farmer demands for tractors sent IHC scrambling, as happened in mid-1955. Certain territories sold high volumes of diesel M and M-TA (Torque-Amplifier) tractors, but IHC ended production of those in October 1954. While Engineering completed preproduction development of the new Farmall 350 Diesel (using a direct-start Continental engine) to introduce in 1956, McCaffrey authorized the Tractor Division to look into acquiring outside-built engines. (R.M. Sheppard, Cummins, and Detroit Diesel eventually provided repowered engines for larger 450-series models.)

McCaffrey's model proliferation occasionally blurred lines between McCormick-Deering-Farmall farm-tractor lines and the International utility models. Throughout the history of the

continued on page 225

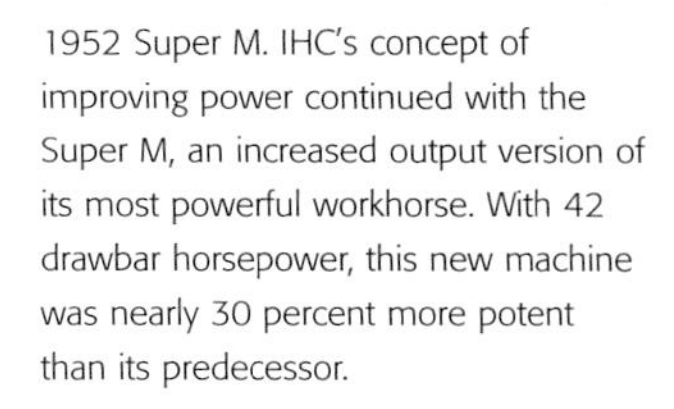

1952 Super M. IHC's concept of improving power continued with the Super M, an increased output version of its most powerful workhorse. With 42 drawbar horsepower, this new machine was nearly 30 percent more potent than its predecessor.

Four cylinders of 4-inch bore and 5.25-inch stroke yielded a 264-cubic-inch engine. Running at 1,450 rpm, the engine would develop 47.5 horsepower on the PTO shaft.

IHC produced Super Ms at both Louisville and Farmall Works. Between 1952 and 1954, the company manufactured 44,551 of these models.

The M and Super M weighed 5,100 pounds with both 6-volt electric starting and lighting systems. The tractor stood 79 inches tall and nearly 135 inches long.

1953 Super B-MD. While IHC produced just about 5,200 of these in the United States, records estimate the British M diesel production at more than 900. Other than country of origin, however, the machines are very similar.

The M-diesel started on gasoline and then once its cylinders reached operating temperatures, the engine ran on diesel fuel. IHC rated both the Super M and Super MD with identical drawbar horsepower (42) and belt-pulley or PTO (47.5) output.

Continued from page 221

Cub, Super C, and Super M tractors, IHC became aware of utility buyers using the solid-axle tractors for farming purposes. This continued with 300 Utility models, the W-4 replacements, to such extent that 87 percent of first-year sales went onto farms. R. W. Dibble, general sales manager, likening these to Ford-Ferguson N-series tractors, encouraged this application and set up dealer strategies to further promote it. Tractor Division added diesel-engine and high-clearance versions to the International 350 lineup. These duplications began to resemble the overlap of Titans and Moguls.

1954 Super H. Nate Byerly of Noblesville, Indiana, worked his Super H with a Little Genius two-bottom twelve plow. IHC manufactured 28,691 Super H and Super HV models in 1953 and 1954.

Steadily decreasing revenues strained budgets. Debt increased from building and equipping factories and acquiring outside resources. These, coupled with a sales-force-driven corporate strategy, forced management to tighten development schedules further. By the late 1950s, this issue arose regularly in EC meetings. Referring to cylinder head difficulties with diesel engines used in tractors, combines, trucks, and power units, R. M. Buzard from National Sales challenged McCaffrey over "the possible impairment of new product development as a result of the demands made on the time of Company personnel to assist in the correction of current problems. Such practices serve to extend current product difficulties to the future and the realization of future sales [is] dependent upon the early introduction of new equipment."

This was the conundrum facing McCaffrey: Diminishing resources forced him to cut corners. His sales background created sympathy for IHC personnel who moved product. He had little

1954 Super M-A. Lurking in the shadow of the right rear wheel were the complicated collection of clutch levers and linkages that engaged the Torque-Amplifier (TA). This system used an auxiliary planetary gear-set to take better advantage of engine torque in difficult conditions.

The Torque-Amplifier essentially doubled the number of gears in the tractor transmission. The planetary gear system allowed operators to shift from one range to another while moving.

From the flywheel forward, it was a pretty simple machine. The 264-cubic-inch inline four-cylinder engine was no different from non-TA-equipped M or Super M models.

IHC produced 26,924 of the Super M-TA models in 1954 only. Rear tread width was adjustable from 59 inches out to 89.

understanding of those who invented or manufactured it. To McCaffrey, when a prototype or two worked, especially when they incorporated proven technology, there was no reason to delay production. Sales, to justify accelerated development requests, began exaggerating sales potential. Manufacturing designed the assembly line and ordered raw materials to meet sales projections and it "priced" the tractor (or truck, combine, crawler, or refrigerator) accordingly. Labor needs were organized prior to assembly.

In February 1956, Mercer Lee, from Finance, gave McCaffrey some sobering figures. "The present production program is in excess of the revised sales estimates by 17,000 trucks and

1954 Super M-TA. David Bradford of Warren, Indiana, kept an eye on the furrow as he plowed with a Model 70 3-14 plow. Engaging the Torque-Amplifier reduced ground speed by about 32 percent while increasing pulling power by nearly 48 percent.

1954 Super W-6-TA Diesel. Bill Tyner's uncommon machine sat waiting for work. The 5,815-pound tractor was one capable worker.

The hefty 264-cubic-inch diesel developed 43.8 horsepower at the drawbar and 48.5 off the PTO. With five speeds forward and the TA doubling that potential, there were few conditions that would trouble this tractor.

The W-6 sat on an 81-inch wheelbase and stretched 130 inches long overall. They measured almost 92 inches to the top of the exhaust pipe.

exceeds current retail sales experience by 15,000 tractors," he said, and "current material stocks were estimated at $26,000,000 in excess of the budget." (That would amount to about $208 million in 2005 dollars.) From a financial perspective, IHC was nearly out of control. Eventually, sales reports would come in two forms: projected and estimated. The former was the number hoped for, the latter was the realistic expectation. Surely somewhere they had a column labeled "actual."

IHC's sales slipped where it made its largest investments. Refrigeration products never reached mainstream urban retail outlets such as Sears, Wards, and Penney's. Company stores were in the country; once farmers had refrigerators and freezers, that market (barely 2 percent of the nation) was saturated. McCaffrey faced becoming a "white goods" maker, offering stoves, sinks, washing machines, and driers, another huge capital investment. A merger of big manufacturers into a conglomerate owned by Sears, Roebuck & Co. (much as IHC was in 1902), claimed most of the business, and McCaffrey unloaded the struggling Refrigeration Division for $19 million in 1955 (equivalent to $152 million in 2005).

Maurice True pulled three 16-inch plow bottoms easily through dry soil in central Indiana. With 5,815 pounds on the ground and the TA providing ten speeds from 1.625 mph up to 16.125, a W-6 squatted down and dug in.

Quality and Performance Lead to Engineering Renaissance

A four-speed transmission gave this compact tractor working range from 1.9 miles per hour up to 12.8. Its offset seating position first appeared on the small Model A and B tractors more than 20 years before this.

Chapter 8

1955–1965

McCaffrey loved construction equipment. Yet agricultural implements, one of IHC's two "core" industries, remained a mystery. Once he authorized manufacture, he reassigned the same engineers who had rushed to finish and release the new products. If customer products failed, the same engineers split their time. They had new projects, but they had to create fixes as well.

McCaffrey acknowledged the need for a proper balance between experimentation and development in Advance Engineering, and Product Engineering in which his staffs prepared items for production and did revisions and redesign afterward. "When a decision has to be made on a Product Engineering program," he wrote to Farm Equipment Division (FED), "it should be expected that enough work will have already been done by Advanced Engineering to provide a sound basis for judging the feasibility of proposed Product Engineering work." For 1956, the ratio was $600,000 for advanced (roughly $4.8 million in 2005 dollars) and $2.5 million ($20 million) for product. Then he cancelled all advanced engineering for 1957.

INTERNATIONAL
140

By the late 1950s, due to rapidly expanded product lines and burgeoning problems, IHC was tilting off balance.

Farm Equipment Division saw the need for big horsepower machinery and worked with IHC's subsidiary, Frank G. Hough, Company, to create this four-wheel-drive prototype in 1959. Hough engineers gave it two- and four-wheel steering and the capability to crab. But it needed much more horsepower.

On May 4, 1951, the Board of Directors had elected McCaffrey its chairman and chief executive officer, presenting its ultimate rebuke to Fowler McCormick. Paul Moulder became IHC's president. In fiscal 1954, gross sales fell from $254 million to $166 million ($2.03 billion to $1.33 billion, adjusted for inflation). Net income plummeted from $10 million to $2.5 million (approximately $80 million to $20 million). Estimates for 1955 hit only $1.7 million (about $13.6 million). Every division cut expenses, labor, inventory, and experiment costs. Despite this austerity, nothing appeared promising to the finance people.

1956 Farmall Model 400 with Model 120A Cotton Picker. Cotton was an important crop for IHC and the company had started developing mechanical cotton harvesters in the 1900s. Beginning with Model H and M tractors, IHC mounted the pickers backwards and the tractors operated in reverse.

The Farmall 400 was the next iteration of the Super M-TA. The new machine provided 51 horsepower at the PTO and ten forward speeds through the Torque-Amplifier, making easy work of any cotton harvest.

In April 1956, the board renamed the Industrial Power Division the Construction Equipment Division (CED). McCaffrey's cherished TD-24, the flagship of his fleet, sold 1,136 in 1953 (at 90 percent factory capacity) and produced $1.8 million ($14.4 million) gross income. For 1955, it reached only 60 percent capacity and 1957 looked no better. EC members proposed combining construction and farm tractor factories. But these ideas evaporated quickly. Tooling was different and retooling was so time-consuming and costly that there was no benefit to shuttering one factory and combining reduced output in another. The only way to achieve real savings was to cut product lines. McCaffrey could not do this yet.

Engineering let outside manufacturers do development. Arnold Johnson, by now vice president of engineering, worked with Frank G. Hough Company of Libertyville, Illinois, on a prototype four-wheel-drive (4WD) farm tractor in late 1956.

1957 Farmall 130 High Clearance. IHC introduced in 1956 the Farmall 130, 130 High Clearance, and the International 130. They remained in production into 1958.

Shortened advance engineering cycles wreaked havoc on IHC's production schedules and reputation: The large turbocharged-diesel 817 engine, in development for five years, still scuffed pistons. Tractor Division was ready to shut down the 650-series Farmall and International production line, possibly replacing the tractor with the new 660-series six-cylinder machine. And CED's ongoing request for a smaller crawler reached financially strapped ears until Tractor Engineering proposed modifications to its proposed Model 340, utilizing 70 percent existing components.

Farm tractors fared no better. The EC cut daily production at Farmall and Louisville by 66 percent over two years. By mid-1957, these regular staff reductions had hammered employee morale. Build-quality deteriorated. The dearth of future product engineering disheartened IHC's engineers and it enabled competitors in second or third place behind them in sales and

IHC manufactured just 1,057 of these compact high-clearance models. Ohio collector Jay Peper found this uncommon original and added it to his fleet.

reputation to tighten the gap. Paul Moulder had no resources available to him. The threats to IHC's position were palpable. Then two things happened, and neither was good for IHC.

First, Deere & Company's board named William Hewitt its chairman in 1955. Soon after he assumed his new job, a new product announcement from IHC's Farm Equipment Division crossed his desk. The product was not important to Hewitt, but the motto across the bottom woke him up. In red print it said "Not Content To Be Runner-Up." He wondered why Deere had been content chasing IHC.

The second event was the rapid failure of Farmall 560 tractors during and after the 1958 fall harvest season. The final-drives couldn't handle the horsepower and torque of their new 60-horsepower six-cylinder gas and diesel engines shoehorned into what was basically a 34-horsepower Model M tractor.

The standard 130 provided 21.875 inches of ground clearance. The raised version added another 6.125 inches for a total of 27.

By the end of 1958, the year IHC introduced 40-series and 60-series tractors in every power range farmers might need, the company sold $391 million ($2.93 billion in 2005 dollars) in farm equipment. Deere sold $464 million ($3.48 billion). IHC was now runner-up.

IHC's board elected Frank Jenks, a former accountant, as company president in 1957 and then to board chairmanship in May 1958, when McCaffrey retired. One of Jenks' initial acts was to fire the chief engineer of the 460 and 560 project. This was misplaced retribution. Those deserving discipline were decision makers who slashed testing budgets and hurried product launches.

continued on page 245

The inline four cylinder with 3.125-inch bore and 4-inch stroke developed 21.1 horsepower at the drawbar and 23.1 off the PTO. The 2,800-pound tractor stood 71.75 inches tall at the top of the steering wheel.

The Farmall 230 developed 19.4 drawbar horsepower and 24.8 off the PTO out of its 123-cubic-inch inline four. Cylinders measured 3.25 inches of bore with a 4-inch stroke. IHC manufactured 7,462 from 1956 into 1958.

1957 Farmall Model 230. The colors are Harvester Red and Harvester White. They were the new 1956 and 1957 color scheme for all U.S. Farmall models from Cub through 650.

1958 Farmall 340. IHC introduced the 340 as a new model range in 1958. These models remained in production into 1963.

Hydra Touch was the latest version of the Touch-Control hydraulic system first developed on Frame-All prototypes. By 1958, this was a very sophisticated system.

Continued from page 239

After several hundred hours, Farmall 460, 560, and International 660 tractors' final drive gears began failing. In early 1959, nearly 4 percent of the big 60-series sat outside dealer service doors. It became an epidemic of the inoperable. Bull gear and pinion sets showed "the tendency toward galling," defined as the "tearing apart of metals due to overexposure to extremely high temperatures as can occur in inadequately lubricated [systems]." Farm Tractor Engineering Department (FTED) wasted no time. It revised bull gears, pinions, and brake shafts for every tractor that was still in production. Then, on April 24, it expanded revisions to include the differential bevel gears. On June 12, it began revising tapered bearings and redesigning the entire differential case.

Tight finances hampered emergency response. Engineering reported it could have new gears manufactured for the 460s by late September 1959, and for 560 and 660 models a month later. Until then, it had no replacements either for tractors in production or for the 3,000 IHC already sold. It had to replace failed sets with identical sets it knew would fail again.

To restore farmers' faith, the EC doubled the warranty on those lines out to 12 months and 1,500 hours on differential and rear-end assemblies. Then, in a report meant to remain confidential, it authorized full-replacement costs and the 19-hour labor charge for the 594 affected 460 models and 428 of the 560 models. When the project was completed (FTED replaced the parts July 20), IHC had spent more than $113,000 (approximately $791,000, adjusted to 2005

1957 Farmall Model 400. These 400-series tractors succeeded the Farmall M and International 6-series models using the same C-264 engine. The 400s provided the independent PTO and the Torque-Amplifier that first appeared on the Super Ms three years earlier.

1957 Farmall 450 Demonstrator. Brass Tacks Demonstrators are hard to find these days. Dealers ordered them just as they had done with the white Model C and Cubs in 1950. But most Brass Tacks got repainted before delivery and owners didn't want the signage.

The 450 used IHC's C-281 four cylinder with 4.125 inches of bore and 5.25 inches of stroke. These beefy engines developed 51.3 drawbar horsepower and 55.3 horsepower on the PTO during their tests at University of Nebraska.

values), on 460 and 560 customer tractors alone. But the report leaked out, especially a paragraph revealing that "the marginal status of the final-drive components on the Farmall 300, 350, 400, and 450 series tractors was also becoming apparent in the field after one, two and three years of service." After this, no recall could save the tractor's reputation.

However, engineers were not nearly finished with the troubles the tractors would give them and their customers. Almost two years into the production run of 560 and 660 diesels, crankshafts began breaking (42 had failed by July 1960). The much higher compression of Increased Horsepower diesels made weaknesses more apparent; this did not occur in gas engines.

But it hadn't ended yet. Customers returned to dealers for the same repair a second and third time. The payback for 1957's unbudgeted advance engineering program cost a fortune. In mid-July 1960 IHC interrupted I-660 production for two weeks to revise all tractors still there. Then, in late October, IHC called back all unsold I-660 models to the factory for disassembly. For each customer tractor, they replaced 19 mandatory and 10 as-inspection-indicated pieces at no cost. This process required 70 hours of labor. It cost IHC nearly $376,000 (some $1.93 million in 2005), not including parts for 1,829 chassis and 1,667 diesel engines. To provide a diesel for small-tractor customers, the board chose to import the Doncaster-built B-275 tractor to U.S. markets. Already available to Canadians, the Standard McCormick International diesel headed to the United States on May 4, 1959.

These big 450s stretched 147 inches long as a wide front (or 4 inches shorter for the row-crop version,) and stood 80 inches at the steering wheel (plus another 12 inches to the top of the nose sign.) They weighed 5,600 pounds, dry.

Nine months later, on February 4, 1960, the Farm Equipment Division approved "Federal Yellow, No. 483-21 or No. 483-23 oven dry or No. 483-22 air dry" paint as standard equipment on all International 340 and 460 Industrial tractors and optional on International 240, 340, and 460 Utility models. IHC concluded it was more visible at night and admitted "yellow coloring appears to create the illusion of a more massive appearance, which is beneficial in the sale of industrial equipment."

On March 31, the division extended that decision to include the Cub (optional), Cub Lo-Boy (standard), International 140 (standard), and International 460, 560, and 660-series industrials (all optional). On standard yellow tractors, "The current red and white color combination will be available optionally when so ordered." (This set of rulings would create havoc among collectors and restorers for decades to follow.)

INTERNATIONAL
80
COMFORT

UNDER FRANK JENKS, the EC moved ahead with the Improved, or Increased Power, 240X (35-horsepower) and 340X (45-horsepower) line of tractors scheduled for production in July 1961. Part of the improvement was final settlement of the three-point hitch dilemma:

"One of the reasons why our tractors in the horsepower category of these models have been losing ground to competition is the lack of an adequate three-point hitch. In order to regain our former position in the tractor sales in this particular category, it will be necessary to have a three-point hitch and hydraulic system equal to or better than the Ferguson system." The EC included these on Improved 240X and 340X tractors tentatively set for July 1961, and on the 460X and 560X models for 1962. The new models, designated the 404 and 504, retained 240- and 340-series styling until introduction in 1962 of the 706 and 806. All four lines appeared in the new bodywork of the large tractors.

Jenks was no stranger to cost consideration but no friend to sales pressure. Even though he had tightened budgets, he still intended to improve products. IHC's oldest factories, the McCormick, Milwaukee, and Rock Falls Works, were inefficient. Because of their construction and layout, IHC could not reconfigure them into the facilities it needed to assemble tractors economically or quickly. Jenks closed the McCormick and Rock Falls plants, leaving his successors to deal with Milwaukee.

Opposite
Indiana farmer and collector Martin Thieme finished up the last of his bean harvest on a club Plow Day. He hooked up his Brass Tacks demonstrator to an International Model 80 bean harvester and gave a crowd of friends a living history lesson.

The good old days meant harvesting in the sun, heat, and dust. That's why many farmers say the "best" good old days are today.

Plow Days events bring friends and collectors from all over a region to one location. It gives owners and enthusiasts a chance to meet, reminisce, share a meal, resurrect old memories and make new ones for the younger members.

At the same time, he quietly enlarged the Engineering staff. He approved experimental projects and demanded more complete testing. After the 60-series disasters, Jenks pushed every new line introduction back a year, because "there would have been an insufficient length of time to do an adequate job of testing." With Engineering's contributions acknowledged at last, projects began to reinvigorate a department that had taken hard criticism for errors forced on them.

In mid-March 1960, International Harvester Experimental Research (IHER) began work on a "full hydrostatic drive research tractor using commercial pump and motor units with I-340 tractor components." E. Jedrzykowski, a staff engineer, described the system in a number of internal reports and outside papers.

"Full hydrostatic drive replaced clutches, spline shafts, axles, gears, etc., used for propelling conventional tractors. The research hydrostatic tractor proved the feasibility of using a single hand

control for infinitely variable ground speed regulation in either direction at constant engine speed with the ability to brake with the same control lever on deceleration." The continuously variable ground speed at steady engine speed allowed PTO-driven attachments full independence. For plowing, rototilling, or snow blowing, with engine speed set for maximum torque, the hydrostatic drive made full engine power available from zero miles per hour up to maximum ground speed. IHER devised the systems in January 1959 and tractor design began in June. Jedrzykowski led the group that installed matching motors for each rear wheel, "used in parallel which eliminates the need for a differential." Jedrzykowski specified 188-cubic-inch radial motors built in England and fed by a variable-displacement pump. Main power for the hydrostatic pump came from an 80-horsepower Solar Industries Titan T62T single-shaft gas regenerator turbine. The tractor, designated the HT-340 (Hydrostatic Turbine), first ran December 7, 1959.

1958 Farmall Model 240. Vercel and Marilyn Bovee's 240 Row Crop shines in the late fall sunlight. IHC kept these models in production from 1958 into 1961, manufacturing a total of 3,710.

Farm Equipment Division model nomenclature got confusing in the late 1950s. These 40-series tractors followed the 50-series models, and they appeared at the same time as the larger models became 460, 560, and 660.

The Hydra Touch hydraulic control system managed elevation and draft control of any implement attached to IHC's three-point hitch. The Traction Control Fast Hitch rapid implement mounting system (holding the drawbar on this tractor) made tool changing quick.

1958 Farmall 350 High Clearance. The skies of western Iowa sometimes get quite dramatic. Bob Pollock's tall tractor reached nearly 100 inches to the top of the exhaust stack.

The hydrostatic drive eliminated the shock upon engaging forward drive under heavy implement load, minimizing strain on drivetrain components. By mid-March 1960, Jedrzykowski and his colleagues found an opposite problem. With no load, or at high speeds (road travel), pressure dropped in the pumps. This affected hand-control response. Hydrostatic braking was nonexistent. IHER added separate brakes purchased from Truck Division. After IHER concluded its development work, it repainted the prototype's sleek fiberglass body created by IHC's chief industrial designer Ted Koeber. For 1962, IHER added a three-point hitch, larger tires, and rear lights; it also stabilized steering, slightly desensitized controls, and transformed the prototype into the HT-341. On September 1, 1967, after years of field tests and demonstrations, and static displays, IHC donated it to the Smithsonian Institution in Washington, D.C.

ELWOOD FOUR-WHEEL-DRIVE CONVERSION KITS soon came to the EC's attention. Farmers who owned later-production Farmall M, 460, and 560 tractors were converting these machines to aftermarket 4WD. The EC reinstated the Frank G. Hough Company's development programs

IHC manufactured just 151 of these tractors. The adjustable front end offered 60-to-90-inch tread width and provided 11 inches more ground and crop clearance than from standard 350 models.

1959 International 660 Wheatland Diesel. This 9,875-pound machine worked effortlessly through dry soil in Central Indiana. With 64.4 horsepower on the drawbar, this tractor felt barely any strain from the five-bottom plow behind it.

on an IHC-produced true 4WD. New chairman Harry O. Bercher (who replaced the retiring Jenks in May 1962) asked IHER to test the Elwood and other kits. IHER found them weaker in durability and performance than a comparable-horsepower, true four-wheel drive.

Throughout this period, small-farm operators continued to sell out to larger neighbors. The U.S. Census projected that the number of farms of 500 to 1,000 acres and those of 1,000 acres or more increased by 5 percent from 1954 through 1959. On February 1, 1962, the EC launched a two-tractor program with 105-drawbar-horsepower diesel engines driving through sliding gear transmissions, to be designated the International 4100 Four-Wheel Drive tractor. Co-developed with its Frank G. Hough subsidiary, the prototypes began testing August 1962.

In late August 1962, FED authorized high-clearance versions of the new 504 Farmall following requests from cane farmers in Mississippi and vegetable farmers in Florida. In October, after introduction of the 404 models, dealers telegraphed to IHC their disappointment "that the tractor did not have a constant running, or independent, type of power take-off such as available on Ford, Massey-Harris, or Deere models in the 35-horsepower category." While the 504 model had it, and the 404 had a transmission-driven PTO, the EC eliminated "live" take-off from 404 design specifications to save money. Cost analysts determined that IHC needed to sell 636 of the 404s with live PTO to repay belated development and tooling costs. Sales estimated the company would sell 1,500 tractors with live PTO that would go to other companies

The 36.8 drawbar horsepower high clearance tractor provided ten forward speeds because of the Torque-Amplifier. Operators could run from slower-than-1.6 miles per hour up to 15.5 on the road in transport gear.

IHC manufactured its 70-series plows between 1958 and 1965. This was the five-bottom 14-inch version. The Wheatland ran 18.4-34 rear tires and 7.50-18 fronts.

if the new tractors didn't have it. Harry Bercher, who shared Frank Jenks' support for and belief in IHC's complete revitalization, approved it.

But this was barely six months into Bercher's job, late in October 1962. FEREC was at work on the Hough-based 4100, 4WD prototype. FED began planning the next large two-wheel-drive tractors, diesel engine only, producing 120 PTO horsepower with 12,000 pounds of drawbar pull.

This four-wheel, non-Farmall-type tractor had a three-point hitch, Torque-Amplifier from the 806, hydraulically actuated 1,000-rpm PTO and a dual-system two-way hydraulic drawbar to push implements down as well as lift them. Engineering began design layout on January 2, 1963. The EC allocated $800,000 (about $5.2 million in 2005 dollars) for design, prototype assembly, and testing. FED had three prototypes ready in September. Originally the EC planned to release this Model 1206 to Production in January 1964; Engineering needed until June 1964. Tight finances were not what delayed the project. One prototype went to the Engineering Center for 2,000-hour-endurance track testing where it showed 98.2 drawbar horsepower while the other two went to Pecos, Texas, for customers' use. In Texas, one 1206 did heavy-duty deep plowing; the other, equipped with four-wheel drive, did ripper and land-leveling work. They were good, powerful tractors, almost too powerful, in fact.

Tire technology held up the 1206. Once Solar Division's turbocharger spun the diesel engine to full power, tire sidewalls began buckling; prototypes peeled lugs off treads and wheels spun on the tire beads. Project engineers from Firestone and Goodyear redesigned belts throughout

1961 Model 140. IHC introduced the 140 series in 1958 as both a Farmall and International model. They remained in production into 1959. The adjustable front axle offered tread width from 44 to 70 inches.

the casings and reoriented lugs to grip as well as unload mud. They developed an 18.4x38 heavy-duty tire specifically for the 1206. Once FEREC had tires that would stay underneath their prototypes, development continued. Production as both a Farmall and International 1206 Turbo began July 1965.

Early in 1964, the EC agreed with FED that there would be benefits in standardizing tractor designs for all of its markets. These two groups birthed the concept of World Wheel Tractors, and its first committee meeting, July 9, 1964, set up development programs for a 40- and 50-PTO-horsepower tractor through joint production in U.S. plants and at Doncaster, England. FEREC, at its Hinsdale facility, designed the tractors, and by the second meeting, April 15, 1965, the large tractor had increased to 52 horsepower. FED decided soon after to manufacture complete tractors for IHC's overseas markets at the Doncaster Works, while Louisville Works would complete 17 varieties of partially completed tractors for delivery on skids. IHC introduced these as the "International" 454 (40-horsepower farm-utility), 2454 (industrial-utility), 574 (52-horsepower farm-utility), 2574 (industrial-utility), and 574 (row crop).

In a blurring of the lines, Doncaster supplied diesel 454s to Canada, while Louisville provided all other models to Canada. Gas engines came from Louisville, while IHC's Neuss Works

The 40-Series. At Keith Feldman's farm, it was a gathering of the 40-series row-crop family. The 240 is at left, the 340 in the rear, and the 140 is in the foreground.

in Germany produced diesels. FED offered tractors either with mechanical or hydrostatic transmissions. FEREC designed bodywork for the series to create a "family styling" silhouette and appearance.

The trend toward larger farms needing more powerful tractors became clear to FED product planners when they read responses from 615 farmers in early 1963 about their future tractor wants and needs. Fifty percent of the respondents ran farms larger than 500 acres, and 30 percent claimed gross sales beyond $40,000 the previous year (approximately $260,000 in 2005). They averaged four tractors per farm (although a few collectors admitted to having more than 20 by then). Only 11 percent owned tractors with more than 70 horsepower in early 1963, but 29 percent felt they would need that power in five years. On farms larger than 1,000 acres, 58 percent needed 4WD, either below 65 horsepower or with 95 horsepower or more. Gasoline was still their fuel of choice for up to 50 horsepower; between 30 and 70 horsepower, owners preferred LPG, while users above 60 horsepower wanted diesels. Sixty percent of the farmers mounted plows rather than pulling them.

While its Diesel engine made this a desirable "collectible" tractor, its history as a Demonstrator added to that luster. But it was the Buddy Seat that allowed the dealer to show a prospective customer the tractor capabilities from the operating platform that made this a rare and desirable piece of Farmall history.

1961 Farmall 560 Diesel Demonstrator
Wilson Gatewood's demonstrator model showed off its gold-painted four-bottom Super Chief plow. IHC introduced the 560 in 1958 and kept these nearly 60-drawbar horsepower machines in production into 1963.

1961 McCormick International B-275 Diesel. This utility model plugged a hole in IHC's tractor line in the United States. With 30.9 drawbar horsepower and a three-point hitch with mechanical weight transfer, live hydraulics and an independent PTO, the Doncaster import became quite popular in The Colonies.

The survey also told IHC that owners of tractors with less than 40 horsepower were more satisfied with their machines than those with more power. "This may suggest," the unidentified researchers concluded, "that the demand of tractors is dividing into two classes: small tractors and large tractors." For IHC, whose sales had jumped, thanks to 1964 and 1965 tax cuts, this was not bad news.

Bercher invested in his business, pouring money like molten ore into Wisconsin Steel, IHC's boutique mill, and even more into McCaffrey's favorite, the Construction Equipment Division. He moved the company from its long-time world headquarters at 180 N. Michigan Avenue north across the Chicago River into the brand-new Equitable Building at 401 N. Michigan, on the site of Cyrus McCormick's first reaper factory.

IHC, with its varied interests in steel, construction, trucks, and agricultural equipment, impressed customers and Wall Street investors as a corporation gone to greed: If it could do well in this, it should try that too. Still, with more successes than failures like 560s and 660s, the customers and the stock market remained loyal. However, IHC's resources, grown fat in the wealthy mid-1960s, were not limitless. It soon found itself laying too thin a financial blanket over its core industries.

To give small tractor sales a shot in the arm, Bercher agreed to restyle Farmall and International 140 models and the Farmall Cub and International Cub Lo-Boy in January 1963, to match 706 and 806 tractors. FED proposed an International 606 model to replace its long-troubled 460, ending production in April 1963. Manufacture began in March 1964.

1961 Model 4300 4WD. This machine resulted from experimental efforts with IHC subsidiary Frank G. Hough. Hough created the 4-WD-1 in 1959. IHC's directors felt it needed much more power and Hough followed up with two more prototypes. The unit called 4-WD-3 went into production as this Model 4300.

Iowa IHC implement and tractor dealer Jerry Mez has built an exceptional collection of IHC's history. This powerhouse is just one of the treasures. The Model 4300s ran on IHC's D817 turbocharged inline six-cylinder diesel that developed 214 drawbar horsepower.

John Wagner's handsomely restored Doncaster-built tractor enjoyed one last sunny fall afternoon before consignment to winter storage. The British 450s weighed 6,877 pounds, measured 141.125 inches long and stood 80.75 inches to the top of the steering wheel.

1962 McCormick International B-450 Diesel. It looks like a United States-produced Model 400 with its single color scheme. Manufactured at IHC's English plant at Doncaster, its engine is the same, as well. The company's Australian operations used this D-264 inline four-cylinder diesel in their models too.

The diesel developed 43.8 horsepower on the drawbar and 48.5 horsepower off the PTO shaft at 1,450 rpm. The familiar Torque-Amplifier offered ten forward speeds ranging from less than 1.7 miles per hour to more than 16.5.

Responding to increasing requests, and rejecting the Elwood system, FED produced its own four-wheel-drive tractor using a front axle developed by American Coleman Company with a differential from current production at Fort Wayne, Indiana, Truck Works. Because of its size, Farmall Works completed assembly on 706 and 806 4WD models in its Special Feature Department, established to accommodate small-series production requests. IHC offered the 4WD option with the start of regular production in August 1963.

Two years after approving development of the 120-horsepower 4WD, to be known as the Model 4300, FED shelved plans for the smaller 4100 4WD codeveloped with Hough. While everyone agreed the tractor had potential, Deere introduced its 5010 with 109 horsepower and others had 4WDs closer in power to IHC's 4300.

To finish 1963, FED proposed blending its International 404 with British B-414 models to make the 424. It used the B-414's diesel engine with an independent PTO while absorbing the I-404's swept-back front axle and its better looks. This Anglo-American hybrid accepted the full range of accessories industrial customers wanted, taken from domestic parts bins without the need to import the modifications. FED determined the 424 would be production-ready in nine months, around May 1964, whereas engineering from scratch required three to four years from proposal to first sale.

1958 Model 460 with International. Model 234 Mounted two-row Corn Picker. IHC introduced this mounted harvesting unit in 1971. Implement engineers designed it to pick, shell, crack and even grind the corn before delivering it to the wagon towed behind the tractor.

The International 606, approved in February 1963, entered catalogs as the Model 656 in both International and Farmall colors. Economic constraints combined with new-product teething difficulties to delay development. FED originally planned it for gas, LPG, or diesel, Torque-Amplifier, "Power Shift" PTO for either 540- or 1,000-rpm use, and two- or three-point hitches. The 656 used the hydrostatic transmission and prototypes tested the new XCF-65, a new experimental "cab-forward configuration." FED completed 10 tractors for "acceptance tests" in March 1965. "Cab forward" addressed complaints IHC market researchers had heard about operator seat comfort and access, and inadequate standing position. The XCF-65 was a "low silhouette, deluxe agricultural and industrial vehicle." To provide a tractor with the greatest-possible operator's comfort, the cab-forward configuration improved controls, and situated the cab ahead of the rear wheels with a midship "walk-in" accessibility. The fuel tank went behind the operator's seat, which FEREC moved forward 14 inches from the regular I-606 position to provide for easier entry and an improved ride. The insulated cab provided opening windows, a heater/defroster, and air conditioning.

Two weeks after releasing the 656 to production, FED discontinued manufacturing the Farmall Cub on May 25, 1964, observing that as "the trend towards consolidation of small farms into larger acreages and the fact that less human labor is available to operate these larger farms, agricultural use of this size tractor has declined." FED continued the International Cub, in yellow-and-white only.

As his wagon is filling, the farmer slows to single-row operation with his mounted Model 234 corn picker. Engineers designed this unit to fit IHC Farmalls and tractors from several other manufacturers as well.

IHER constantly sought new uses for new technology. In mid-June 1964, they suggested replacing the existing gear transmission of the Cub cadet. While other tractors in this market used belt drives, IHER suggested IHC leap ahead with a hydrostatic transaxle to eliminate the clutch, transmission, axle, differential, and brake, and reduce tractor weight by 55 pounds in the bargain. They completed a prototype on June 7, 1963, and tested it intermittently through the end of the year, concluding that "the Cub Cadet hydrostatic transaxle should be only the first of an expanding family of hydraulic components for use in vehicle transmissions or as hydraulic power sources for power steering, front end loaders, bulldozers, etc." They followed it up with development plans for hydrostatic transmissions for 504, 606, 656, 706, 806, and 1206 agricultural.

Economic conditions forced a hiatus on continued development of some tractor projects during late 1962 and 1963. The EC delayed some testing and cancelled other projects outright. The Hough-designed International 4100, shelved in September 1963 because of costs and inadequate power output, had been popular in Engineering. Testing and development funds materialized, and by November 1964, preliminary structural and durability tests had only reinforced FED's belief in the tractor. Only a 2,000-hour endurance test remained and manufacture was scheduled to begin after that test, with sales beginning in August 1965.

For Engineering, once again, there seemed reason to have hope.

Sales and Marketing Reassert Their Influence

Chapter 9

1965–1975

During the 1960s, Farm Equipment Division started asking questions. A new market research group anonymously polled recent IHC equipment buyers and registered owners of competing makes. IHC hoped to learn where it stood against the competition, conscious of problems in the past. The truth hurt.

Surveys expressed dissatisfaction, not with the equipment so much as with "the personality of the dealer," which FED's chief market researcher M. J. Steitz took to mean the professionalism and helpfulness to past, current, and future customers. Equipment drove some long-time customers away; recurring problems with 560s turned some families off, even if they had farmed with IHC since the early 1920s. Surveys urged Engineering to improve quality, and Manufacturing to eliminate production defects, but mostly they stressed the importance to Sales of better attention to farmers' needs and situations.

Other surveys allowed FEREC to glimpse the future from the farmer's point of view. These surveys asked dealers to describe the IHC's tractors based on their own local markets.

1969 International Harvester Company of Australia A-564. Just like the Doncaster-built B-450 models, this Geelong product made use of the gas engine cylinder block that IHC used in its U.S. production models. With 48.5 horsepower on the PTO shaft, this was a potent machine.

INTERNATIONAL
564

Most of them saw a trend toward higher-speed plowing; their farmers preferred working at 5 miles per hour with five bottoms instead of 4 miles per hour with six bottoms. Farmers already were requesting tractors with 140 to 150 PTO horsepower. Many wanted dual rear tires, not only to get power to the ground without slippage but also to decrease soil compaction caused by single tires. Dealers told FED they were ready for the next Farmall. They hoped that IHC was ready to start leading again.

FEREC was moving forward. It fitted its first experimental hydro-mechanical transmission (designed back in mid-1961, built by hand, and bench tested) to a regular-production Farmall

1966 Farmall 1206 Turbo. This was IHC's first U.S. turbocharged diesel engine and its first to exceed 100 horsepower in a two-wheel-drive tractor platform.

The 361-cubic-inch six made use of cylinders with 4.125-inch bore and 4.25-inch stroke. At Nebraska, this turbo diesel developed 112.6 horsepower off the PTO shaft.

806 early in 1965. By February 1966, FEREC released it for production, incorporating an "exhaust gas ejector/heat exchanger" to cool the hydro-mechanical transmission oil. This worked similarly to the heat pump concept and offered the added benefit of muffling the exhaust noise without need for a muffler. However, the EC subsequently cancelled its plans and IHC never produced the 806 Hydro.

On March 10, the EC released the F-656 and I-2656 models to production beginning October 1, followed almost immediately by the F-656 Hi-Clear. By May, the EC pushed back its production plans for increased-power, restyled 706, 806, and 1206 tractors from November 1, 1966, to June 1967. IHC sold every 706 it manufactured and could scarcely afford to shut down the line to retool for its replacement. FED completed pilot models of the new 756, 856, and 1256 on September 1, 1967, and began production in October.

Self-examination continued. In May 1966, the CED studied how IHC engines fit its needs and whether its engines applied to farm tractors as well. CED found its lineup from 282 to 429 cubic inches had few interchangeable parts and no family resemblance. While FED used engines by the hundreds or thousands, CED had runs of 250 or 300 scrapers or crawlers; engine manufacture on this scale was too costly.

CED knew that Neuss Works in Germany, despite its modern assembly facilities, made no engines for America's farmers' needs into the 1970s. Two engine "families" grew out of this research: a 300-series, with two engines of 312 and 360 cubic inches, and the 400-series, with three blocks of 414, 436, and 466 cubic inches. To reduce costs, CED minimized the number

A grader blade was an easy load for the 1206. At Nebraska, the tractor pulled a 10,770-pound load at 3 miles per hour.

of engine blocks and maximized the number of parts common to each engine to provide the greatest benefits to cost, design, engineering, testing, and maintenance. CED proposed to Motor Truck, Farm Equipment, and to its Power Unit engineers that they design future products to use these five displacements. CED configured the large 360- and all the 400-series engines for production with and without turbochargers. This offered a power range from 94 to 231 gross horsepower.

In February 1967, FED completed a product plan that addressed dealer and farmer comments. Every tractor, from current-production Cub Cadets up the line to 1206s and 4100s

(introduced in 1966), gained horsepower. FED also planned a new 1556 (with 140 PTO horsepower) and a 4256 (at 160) for production in late 1970. Engineers enlarged fuel tanks to allow longer working hours common on larger farms and configured additional mounting positions to fit tanks onto tractors for those who sprayed chemicals during tillage operations.

A. O. Smith Corporation, an Ionia, Michigan-based company that made fiberglass car bodies and the tops for IHC's Scout utility vehicle, contracted with IHC in March 1967 to manufacture fiberglass tractor cabs based on the survey information. Smith's cabs from FED designs insulated operators against weather, dust, and noise, and they provided air conditioning, heat

1968 Farmall 756. The 56-series introduced operator comfort with hydraulic seat suspension on a 38-inch-wide operator platform.

Gas engine models used IHC's C-291 inline six with 76.5 horsepower off the PTO shaft. The tractor weighed 9,483 pounds and sold new for $10,710.

Hydrostatic power steering with a five-position tilt-steering wheel and a standard 12-speed transmission (offering eight forward and four in reverse,) made operating the 756 a delight.

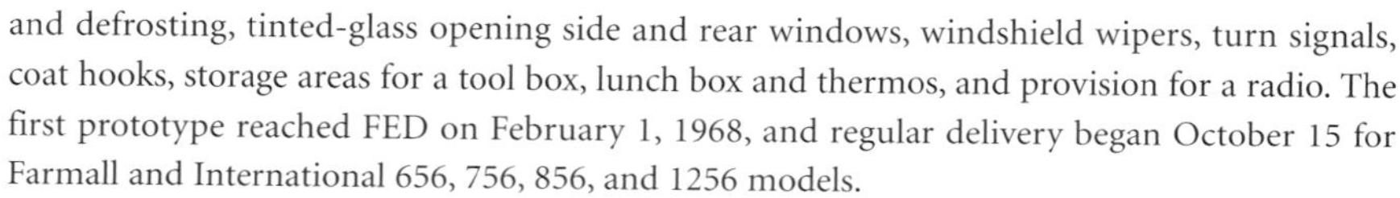

and defrosting, tinted-glass opening side and rear windows, windshield wipers, turn signals, coat hooks, storage areas for a tool box, lunch box and thermos, and provision for a radio. The first prototype reached FED on February 1, 1968, and regular delivery began October 15 for Farmall and International 656, 756, 856, and 1256 models.

As an interim measure, IHC also developed a protective frame that attached to the rear axle carrier to address growing concerns over tractor rollover accidents. This incorporated a fiberglass canopy and seat belt for factory application to 656, 756, 856, and 1256 models beginning in December 1967. FEREC modified fenders and exhaust pipes that were compatible with the new roll-over-protection system (ROPS).

The 806–856 and 1206–1256 models, as well as the 4100, also benefited from a new Category II and III Quick-Hitch coupler FED introduced for use with the three-point hitches. As with IHC's previous Quick-Attach and other rapid-implement mounting systems, it sought to make implement attachment possible without requiring the operator to leave the cab (except for PTO attachment). Compatibility with other makers' quick-couple implements was another objective.

IHC offered an optional two-post Roll-Over Protection System (ROPS) with a canopy for these tractors. Not many buyers went for that option.

In late March 1967, FED solidified plans for its World Wheel Tractor series. The first, a 40-horsepower medium-duty utility-type model for farm or industrial uses, was coded TX-19 for internal documents, and became the International 454. Plans also included the TX-36 in International and Farmall 574 model designations, produced in Louisville for the United States and Canada, for introduction in 1970.

Beginning in 1967, FEREC used computers for experimentation and testing systems. This shortened development time considerably. Back in 1954, Tractor Engineering made its first serious attempt to program a digital computer for farm-equipment engineering problems. With the experiences gained while redesigning 460/560 rear ends and transmissions, the department created a base of knowledge. Now the science of metallurgy interfaced with a body of field experience, and everything from compression spring design to V-belt life projection to rockshaft arm stress analysis to gear design and axle-bearing load assessment was possible in the computer.

1969 Farmall 656 with Disc. IHC offered the 656 with Diesel or gasoline power. This gas engine version developed 44.7 drawbar horsepower at 75 percent load in its tests at University of Nebraska.

BY FEBRUARY 1968, 12 PROTOTYPE World Wheel Tractors—six each of 40- and 52-PTO horsepower units—had completed thousands of hours of testing. One 454 (40-horsepower model) ran 1,550 hours at 125 percent of rated load; a hydrostatic transmission–equipped prototype underwent a similar amount of endurance testing on a tractor dynamometer. One single powertrain had 2,100 hours on it. FEREC released designs to Doncaster, and the British plant produced another six prototypes to begin its own testing.

A small problem arose with hydrostatic transmissions affecting forward and reverse control. It involved few 656 models. FED quickly responded with a small spring to secure the pump-servo cylinder-control valve requiring an hour's installation at company expense. Harry Bercher told FED that supporting Engineering was "in the company's best interest."

In May 1968, IHC elected the fourth McCormick, Brooks, as president and chief operating officer. Brooks, 51, was a great-grandson of William McCormick, who was Cyrus Sr.'s business manager and youngest brother. After Brooks graduated from Yale in 1940 with a bachelor of arts in Victorian literature, his Uncle Fowler hired him into the training program. Brooks started at McCormick Works, sold farm equipment in Indiana in 1941, learned manufacturing at Tractor Works, and became superintendent at Melrose Park in 1947 as they tooled up McCaffrey's first

The 656 Gas tractors, in production from 1965 into 1972, sold new for $7,340. They weighed 6,350 pounds.

TD-24s. He sold trucks in Kansas City, and in 1951, he became joint managing director at Doncaster where he conceived the World Tractor.

Brooks observed that IHC derived most of its profits from farm equipment, most of its sales from trucks, and most of its expenses and losses from construction equipment. He championed the World Tractor concept, though not as a crowd pleaser. He saw it as a means to pare costs by simplifying tractor lines, by collaborating on international designs and adopting universal parts.

TO SUPPLEMENT 656 HYDROSTATIC DRIVE TRACTORS, in mid-June 1968 FED proposed new 826 and 1026 models at 84 and 112 PTO horsepower for introduction November 1969. In July it released the 4156 with 140-PTO and 125-drawbar horsepower to replace the 4100 4WD model. To ensure no future problems with this power/workload increase, FEREC upgraded rear-end gears to specifications that worked with the revised 460 and 560 models.

Hydraulics technology had not kept pace with drawbar power. Tractors now could pull more than they could lift. This imbalance had less impact on fieldwork, but it placed risks on farmers driving tractors with raised implements on the roads. Adding hydraulically actuated

1970 Farmall Model 1026 Hydro. Ron Neese's powerful 1026 Hydro waited with its Woods Batwing Model 3180 mower. The mower could cut a 15-foot swath.

implement-mounted wheels helped, but this also encouraged farmers to use even larger implements. FEREC introduced IHC's latest weight-transfer hitch in November 1968. This system advanced Harry Ferguson's geometric A-frame structure; it transferred loads off the rear of the tractor to the steering and drive wheels. Now the hydraulics could raise heavier implements without lifting the front of the tractor. FED offered production units for tractors up to the 1256 models beginning in December 1969.

FED did not upgrade the 1256 itself because it had planned a replacement to keep up with the newest marketing war in farm equipment: the battle over superior horsepower. Deere's new Model 4520, with 120 PTO horsepower, forced FED to raise the ante with a 125-PTO-horsepower model, the 1456, and a later 155-horsepower model to follow. To accommodate the additional power in the 1456, FEREC increased virtually every dimension and strengthened each element of the tractor. Engineers widened gear faces from the transmission to the rear end. They enlarged the radiator, moved it forward, fit a bigger fan, and give it greater clearance. They increased brake disc diameter from 8 inches to 11.375 inches, and enhanced every brake component accordingly. They enlarged rear-axle diameter from 3.25 to 3.5 inches.

IHC introduced Hydrostatic Drive on its 1967 Model 656 tractors. This system provided an infinitely variable travel speed for the tractor and implements while allowing the operator to maintain peak horsepower or torque performance through challenging crops or terrain.

On February 28, Neuss Works informed the EC that Europe needed larger tractors. With no 80-horsepower machine in their lineup, IHC's French and German distributors marketed Deutz, Hanomag, Renault, or Fiat tractors in that range. Under agreement with FED, Neuss Works imported D-310 engines, which it linked to partially synchronized transmissions to produce a prototype X-47, 80-horsepower tractor by November 1969. Regular production as the Model 846 started March 1971.

The 1026 Hydros first appeared in 1969. In Nebraska University tests, the 407 cubic-inch turbo-diesel engines developed 110.7 horsepower at the PTO shaft.

The 1026 H weighed 10,400 pounds. It sold new for $14,970.

Existing tooling, factory production lines, and parts inventories challenged Brooks McCormick's dream of a World Tractor line. While the 400-series and 500-series worked well and were successful, the world had less need for machines larger than 756s, yet these and larger machines sold well in North America. Horsepower wars drove FEREC, while sales fed IHC's treasury. In acknowledgement, the company invested $24 million ($120 million in 2005 dollars) in tooling at the Melrose Park plant to expand annual production to 40,000 of the DT-466 diesel engines. Late in March 1969, FED authorized a two-phase replacement plan for its large tractors. With production beginning November 1, 1970, FED released the 60-series upgrades, offering small increases in power but also a new grille, hood, side panels, and overall appearance.

The World Tractor philosophy created a model for U.S. markets only. The Louisville Works product committee took requests from eastern and southeastern farmers for a 32-horsepower tractor, larger than Cub 14s and 154s, but smaller than 444 and 454 models. Doncaster provided running gear from their B-275 tractor with the front axle, grille, hood, instrument, and steering mechanism from the World TX-19 and TX-36 tractors. Louisville had a 32-PTO-horsepower tractor configured with few

For 1970, IHC painted side panels, fenders, and hoods gold on many of its demonstrators of the tractor lineup. The 407 cubic-inch turbocharged diesel developed 131.8 horsepower at the PTO shaft.

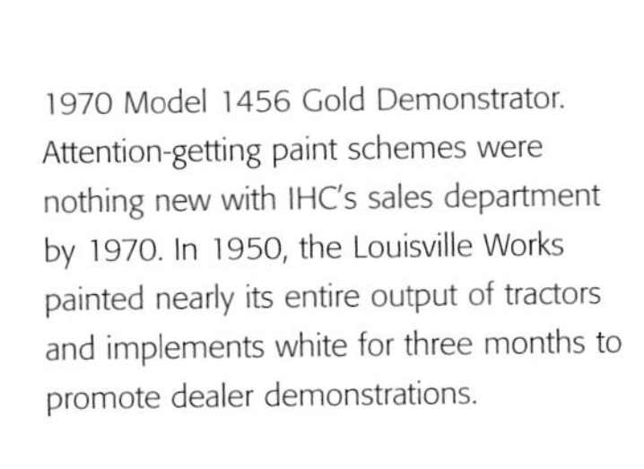

1970 Model 1456 Gold Demonstrator. Attention-getting paint schemes were nothing new with IHC's sales department by 1970. In 1950, the Louisville Works painted nearly its entire output of tractors and implements white for three months to promote dealer demonstrations.

Starting in March 1967, A.O. Smith, a Michigan company that produced fiberglass tops for IHC's Scout utility vehicles, began manufacturing tractor cabs for IHC. The Farm Equipment Division designed the cab to insulate against weather, dust, and noise.

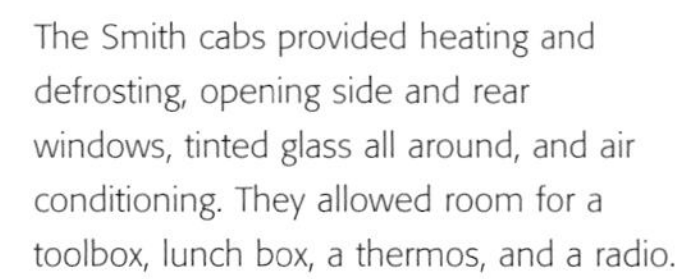

The Smith cabs provided heating and defrosting, opening side and rear windows, tinted glass all around, and air conditioning. They allowed room for a toolbox, lunch box, a thermos, and a radio.

turbo II
Turbo
INTERNATIONAL
1456

1971 International Farmall Model 856 with 1971 International Model 82 Bean Harvester. Matt Thieme harvested soybeans with his 856. IHC produced these tractors from 1967 into 1971. The company's 301 cubic-inch inline six-cylinder gas engine developed 93.2 horsepower off the PTO.

modifications for regional U.S. markets as an International 354. Doncaster did all the testing and development work. FED began production in November 1970.

Yet the Louisville plant experienced problems with the International Cub 154 Lo-Boy. Assembly difficulties caused nearly one in four units to suffer driveline failure. Tight tolerances caused clutch shafts and drive-coupler hub assemblies to wear prematurely. The problem never occurred in hand-fabricated prototypes, but production tooling and fixtures made proper installation very difficult. FED revised the Louisville assembly lines building the Lo-Boys. It also authorized an industrial 25-horsepower companion to the Lo-Boy, using the hydrostatic transmission and a four-cylinder water-cooled Renault engine from France. IHC called it the International 254. Production started in November 1972.

FED had discontinued the 756 diesel from U.S. production in mid-July 1969, replacing it with the 826 using IHC's own D-358 diesel produced at Melrose Park. However, it continued to ship partially assembled 756 models on skids to its plants in France, Mexico, and Australia well into the early 1970s. These plants installed Neuss Works D-310 engines sent directly from Germany.

In late 1970, Mississippi Road Supply (MRS) in Flora began a cooperative program with FED. MRS had manufactured construction equipment and farm implements since 1943. MRS developed and manufactured two 4WD articulated tractors of 130 and 155 PTO horsepower and took over production of the 4156 4WD, four-wheel steering tractor for IHC. In addition, MRS licensed its three current-production 4WD tractors (up to 236 PTO horsepower) to IHC. MRS also gained access to engines using CED's D-466 six-cylinder and Truck's DV-550 V-8 for 130- and 155-horsepower 4WD models. IHC handled worldwide distribution. MRS planned production to begin in February 1971 following a $12,000 tooling expense (about $50,000 adjusted for 2005 values), to adapt IHC engines to MRS chassis mounts. But the program

1971 International Farmall Model 856 with International Model 710 5-16 Plow. Out of the factory, these 856 tractors weighed 8,620 pounds. Adding a little weight to the front end and slipping a dual rear tire on the land side meant that very little would interrupt the progress of this machine.

The 856 engine provided nearly 90 horsepower at the drawbar and 100 off the PTO shaft. With its standard Torque-Amplifier, even six big plows were no challenge.

Well-known tractor-parts supplier Robert Off worked his 856 in hard dry soil. With just five plows behind him, Off preferred plowing faster to plowing harder.

collapsed. IHC then went to Steiger in Fargo, North Dakota, after proposing it use IHC's engines in its large 4WDs.

Big power drove product development, sometimes from outside. FEREC engineers knew that tractor owners modified or replaced IHC's engines to get more power. V-8 engines began to appear not only for power and smooth operation but to give farmers bragging rights, as occurred when Ford tractor owners 10 years earlier installed flathead V-8s for power and prestige.

In June 1970, FED authorized production on its DV-550, 130-PTO (at 2,400 rpm) horsepower tractor. The V-8 appeared in mid-October 1971 as the 1468 model, while keeping the 1466 inline six (DT-436) in production. Two years later, a higher-speed, 2,600-rpm version increased DV-550 output to 145 horsepower.

WEEKS BEFORE BECOMING IHC CHAIRMAN in May 1971, Brooks McCormick's grand plan for simplifying product lines suffered two quick hits. The FED, and its Hinsdale engineering research center, FEREC, in particular, had enjoyed an inspired burst of invention and creativity while Frank Jenks and Harry Bercher tried to get IHC back in balance.

The Sales department, John McCaffrey's source of inspiration, had always rushed Engineering and Manufacturing. Jenks and Bercher together diminished Sales' influence, giving

Starting a new row, the dirt went flying. Off and a friend plowed through bean stubble in central Indiana in mid-October.

Opposite
Twelve suitcases! These weigh something like 75 pounds each. The additional weight up front here went a long way toward maintaining front wheel steering under heavy plow load.

engineers time to get caught up and think ahead. IHC's profitability from a growing economy reinvigorated sales, however. During Bercher's last year, from May 1970 to 1971, even as the economy contracted, Sales resumed its old ways. If any competitor introduced something IHC did not have, it created a vacuum in IHC's line that product planners abhorred.

Describing the International 644, FED's tractor planners proposed fitting the Neuss D-239 diesel engine into a combination of International 544 and 656 components designated the International 654. "The resultant tractor," they said on May 11, "would be targeted to compete with the utility-type Ford 5000 and Massey-Ferguson 175 diesel tractors in this size and price range." Two days later, to fill a gap between the present 154 Lo-Boy and the forthcoming 354 tractor, Sales proposed importing the Kimco 242 tractor as the International 242. Kimco was Komatsu-International Manufacturing Company, a joint venture between Komatsu and IHC created in the late 1960s. "Importation of tractors from abroad has steadily increased in recent years . . . . Figures show 3,220 Japanese tractors imported into the U.S. in 1970. This [242] will place IHC in a competitive position to enter that segment of the market now enjoyed by these imported Japanese tractors." Sales chased a market for 3,000 tractors to provide products for dealers who wanted to carry Japanese imports. Shortly after FED had approved these programs, Brooks McCormick took over.

He inherited a company in trouble. The corporation's entire profits, $45.2 million ($206 million in 2005 dollars), just met shareholders dividend disbursements. Brooks had no reserve

1972 International 1468. From this view, it was deceptively innocent. With the other poles in this farmyard vying for attention, the two big chrome diesel exhaust pipes almost got lost.

It was not easy making it fit. This was 550 cubic inches. It came from IHC's truck division and Farm Equipment Division had to modify it to fit between 25-inch frame rails.

This was one of the most desirable and exclusive of all IHC tractor office seats. If you had known enough to buy one in 1971 or 1972, the tractor cost you just $20,430.

funds. Budgets set for 1972 and beyond had no flexibility. His predecessors closed plants that could not manufacture their products profitably; he inherited employee wage and benefits packages that were more generous and costly than either Deere or Caterpillar. He obsessed over costs and making IHC "well managed." He hired outside executives, hoping to bring in new ideas. Senior managers felt betrayed because they didn't get promotions that, before Brooks, were part of their job description.

Some issues demanded attention. The Vietnam War started an inflation that raised the consumer price index 60 percent from 1967 to 1975. Part of that surge came in February 1971 when the six-country Organization of Petroleum Exporting Countries (OPEC) agreed with 23 of the world's oil companies to a settlement nearly tripling fuel prices in the United States. Oil-producing countries, wealthy from sales of their natural resources, purchased U.S. produce. The Soviet Union bought $136 million ($620 million in 2005) in grain. Famines starved millions in Asia and Africa, which frightened the rest of the world of adequate supplies of food. Farmers expanded their holdings and bought equipment. IHC engineers James D. Wilkins and Richard N. Coleman spoke to the SAE (Society of Automotive Engineers) in mid-September 1971 and addressed one effect of this cycle:

"As the American farmer continues to substitute capital and technology for labor, his need for more powerful tractors will increase. Utilization of

continued on page 296

Ben Warren, IHC's newly named president of the Agricultural Equipment Group asked Farm Equipment Division if they could install this engine in one of their tractors. The rest was history.

Wilson Gatewood's V-8 developed 145.4 horsepower off the PTO shaft. This was an engine that should boast chrome valve covers and exhaust shields.

1468
FARMALL

1975 International 766 Turbo Diesel. IHC manufactured these tractors between 1971 and 1976. Weighing 9,538 pounds, these diesels sold new for $15,240.

Continued from page 292

this increased power can be achieved by the use of tools of greater working widths, performance of several combined operations in one pass, and increasing the field speed at which various operations are performed."

There was justification for bigger, faster tractors. Within nine months of the launch of the 4166 and initial production of the Steiger-4WD tractors, FED had already heard questions about maneuverability and soil compaction hounding FEREC engineers. In response, they mated two 1066 final drives together using a 4166 transmission and transfer case. This articulated 4WD prototype was a forerunner of the series later known as 2+2 tractors.

The Steiger relationship proved that "badge engineering" was convenient, and for the short run, more cost-efficient than IHC spending millions developing their own tractors. Badge engineering involves one company purchasing already-engineered and developed products from another and then putting its own name or badge on the item. This was not a new concept;

The D-360 inline six-cylinder developed a robust 85.4 horsepower off the PTO. The TA system provided 16 speeds forward and 8 in reverse.

McCormick Harvester Company had already done it in the 1890s when Robert Hall McCormick acquired outside-manufactured steam-traction engines, which he sold as McCormicks. Steiger agreed to build a 175-gross-horsepower unit (based on the twin 1066s) and a 275-horsepower version using the DT-466 engine. Steiger would design and manufacture the transfer case, buying transmissions from Fuller. FED designated this as the TX-111.

On April 24, 1972, FEREC introduced its synchromesh transmission to replace the previous sliding-gear type, especially for the 700-, 900-, 1000-, and 1400-series tractors. This four-speed unit provided shift-on-the-go capability under unloaded conditions. It could not accomplish the moving shifts under full load that the Torque-Amplifier could do, nor did it provide the variable-speed characteristics of the hydrostatic. Still, IHC was the only maker offering synchromesh with an optional Torque-Amplifier. Production began October 1973, for the 1566 and 1568 models.

FED released plans for the DT-466-engined 175 PTO-horsepower 1866 and DV-550-engined 1868 models. Based on the 1466 frame, it incorporated a modified planetary final drive from the 1566 and 1568 models and a three-speed sliding-gear transmission instead of the usual four-speed version. Production of this 1500 series began November 1973 and the 1800s started November 1974.

Farmers noticed a change on the side of these new machines. These and all IHC farm tractors released in late 1973 as 1974 model year production no longer bore the name Farmall. Over the previous two years, as the EC eliminated duplicate product lines, the International name grew larger on FED tractors while Farmall shrunk. At the end of 1973, it disappeared altogether.

The engine for these standards was IHC's turbocharged D-414 incline six. The workhorse developed 125.6 horsepower on the PTO.

After six months of discussing consumer products among FED product planners, they were confident that IHC's name was back in public awareness. That faith prompted new efforts to reduce the time and funding that FEREC required for testing new products. Sales had regained much of its former stature. Product Development again assumed a catch-up role, where engineers produced IHC's version of someone else's improvement. It had been half a century since Bert Benjamin convinced Alex Legge to build a few more prototypes to test for another year. Now products were conceived one day and for sale 18 months later.

IHC's engineers continued to innovate. FED and CED encouraged them to promote their accomplishments among their peers. In mid-September 1972, FEREC engineer George F. Boltz addressed colleagues at SAE's national convention, explaining how IHC accelerated tractor testing. During the 1950s, endurance testing went on at Hinsdale's 1.125-mile test track, towing other tractors with engines replaced by water brakes. FEREC needed six operators per tractor to run tests all day. Boltz described days when six or seven prototypes ran at once. FED devised a "tractor treadmill" during prototype testing for the 806 model that rested the 806 on the rear wheels of a water brake tractor set in a pit. The prototype's rubber tires ran against old IHC steel wheels onto which FEREC had welded a 24-inch-wide steel plate. The second technique involved a "tether test," in which a tractor was attached to a 110-foot cable and, with two load machines towed behind it, was set free to go in circles. By running without operators, FEREC routinely accomplished 140-hour weeks on prototype tractors. A thousand hours of testing took just six weeks.

1975 International 1066. Gary Schmitt ran his 1066 with a Parker 450 wagon in front of his 1989 Model-1660 Axial Combine, picking corn in central Indiana in late October. These "black stripe" Internationals were reliable workers that have begun to draw attention from savvy collectors.

When FED introduced hydrostatic transmissions, it kept drawbar horsepower as close as possible to Gear Drive models so buyers could compare work potential. Engine developments increased horsepower, but the Gear Drive's greater efficiency produced higher power readings. With the 66-series, the discrepancy grew so much that 1066 hydrostatic drawbar horsepower fell nearer to 966 Gear Drive statistics. In March 1973, FED renamed the hydrostatic tractors to obscure direct comparisons. It eliminated both 966 and 1066 models, replacing them with the Hydro 100. The 666 became the Hydro 70.

In October 1973, OPEC shut off its wells. That faraway act had ramifications that crashed down around Brooks McCormick's head, making tractor fuel economy a factor in new purchases.

Sixteen forward gears gave the 11,860-pound machine plenty of flexibility, even when towing a full corn wagon through stubble. The tractors sold for $21,080 new.

He had many other serious distractions much closer to home. The Occupational Safety and Health Administration (OSHA) was formed as part of the U.S. Department of Labor in December 1970. OSHA affected tractor operators in the field and engineers creating tractors for them; it also meant that IHC's foundries, manufacturing plants, its Wisconsin Steel Mill, and its test facilities had to comply with strict regulations, no matter what it cost the owners. The previous July, Washington dealt an equally serious challenge to heavy manufacturers with creation of the Environmental Protection Agency (EPA), formed to give teeth to the 1963 Clean Air and Water Act. McCormick saw IHC's aging plants and its dirty, inefficient, and unprofitable foundries and steel operations as future money pits, facing cleanup and modernization costs of hundreds of millions.

Gary Schmitt and his son Isaac maneuvered the Parker wagon. When it was close enough, Schmitt augured his load into the trailer of his 1981 International Transtar cab-over semi-tractor.

IHC's profits increased in farm equipment. In 1975, sales topped $2.1 billion ($8.4 billion, adjusted to 2005 values), and for the first time in 10 years, farm equipment outsold trucks. That was a hollow distinction; the country had endured the worst of OPEC's fuel crisis through early 1974, and the Truck Division experienced a huge loss through 1975. A year later, the Construction Equipment Division lost $4.7 million ($18.8 million). IHC's credit rating dropped to "bad risk" status. McCormick discussed closing CED. However, engine manufacture was tied to truck, tractor, and construction needs. Shuttering CED meant layoffs in the foundries as well. Estimates to pay off creditors and employee separation settlements exceeded $300 million (approximately $1.2 billion), just to shut the doors. The outlook was bleak.

The Stumbling Giant Faces Dissolution and Accepts Rescue

These MFWD models weighed 12,460 pounds and Case published their list price at $59,875. The company offered a High-Clearance variant.

Chapter 10

1976–1999

The country's bicentennial put most of the United States in good cheer. In a patriotic flurry, FED released the 4568, its 300-horsepower tractor built on a Steiger chassis. IHC sold it only in 1976, renaming the big articulated four-wheel drive the 4586 for 1977.

IHC launched the 86-series in November 1976, with two Hydro models, the 86 and 186, and a full range of small-to-large tractors from the International 868 up to 1586, 4186, and 4386. These tractors adopted the A. O. Smith XCF-65 "pod" type cab that moved the operator forward and made room for the fuel tank at the back. It also introduced a 284 from Kimco, the joint venture operation with Komatsu. Along with more power, the 86-series bought farmers the "Control Center," the new weather- and sound-insulated cab derived from owner surveys conducted in the 1970s. The Control Center provided the farmer with more instrumentation than earlier tractors. The polyfoam and iso-mount insulators, thick carpet, and wraparound glass isolated farmers from sounds that typically told them whether all was well with their machines. Dealers taught farmers to trust the gauges instead of their ears.

Steiger developed an even larger chassis for 1979, the 4786, with 350 horsepower. IHC then began to replace the 54-series and the 74-series line in 1980 with the full-diesel line of 84-series small- to medium-sized tractors, reducing two lines into one. These tractors, while not carrying on the Control Center cab, continued some features and introduced hydrostatic power steering, hydraulic disc brakes, and a differential lock along with an optional German ZF-built mechanical front-wheel-drive (MFWD) attachment to make two-wheel-drive tractors into four-wheel drives. The 84-series also provided torsion-bar draft control, planetary final drives, and the three-lever hydraulic hitch control. International 884s came with Torque-Amplifier standard, while IHC offered it optionally on 584, 684, and 784 models.

Mitsubishi manufactured IHC's smaller tractors. These small diesels provided between 15.2 and 21 PTO horsepower as models 234, 244, and 254, manufactured in Japan with IHC-specified features.

1975 International 4366. IHC manufactured these 4-wheel-drive models from 1973 into 1976. Frank Ferguson's tractor worked with the company's 12-foot front blade.

This machine ran dual 18.4-38s all around. Having just completed some road building around the farm, this machine was ready for the snows of winter.

IHC's D-466 engine powered the 4366. The in-line six-cylinder turbo diesel developed 163.1 horsepower off the PTO shaft.

The tractor sold for $37,400 in 1976, without dual wheels or plow. In that configuration, it weighed 18,800 pounds.

FEREC, which had mated rear ends of two 1066 production tractors to create a prototype articulated 4WD, tried the same thing with 86-series prototypes in a continuing effort to develop mid-power-range models. It designated these as 2+2s because they consisted essentially of one two-wheel-drive tractor plus another one, a 1970s version of the tandem tractors on the late 1950s and 1960s. FEREC carried over the Control Center cab system but, unlike Steiger and Versatile, it was unconvinced the cab belonged in the middle. To use the cab and the 86-series final drive, IHC designed the 2+2 so the solid front axle steered by pivoting the front half of the tractor rather than turning individual tires. The long nose housed the engine-before-drive-axle configuration of 66- and 86-series tractors. This required no engine redesign and very little drivetrain modification. It also increased tractor stability with heavy rear-mounted implements.

Before IHC introduced its 84-series in 1980, it brought out the new 3388 (130-PTO-horsepower) and 3588 (150-horsepower) 2+2 tractor in very late 1978. Then, a year later, it introduced the 170-horsepower 3788 2+2, using the latest DT-466B turbo diesel.

Several philosophies emerged from World Headquarters as the bicentennial passed. In 1975, McCormick approached Booz, Allen & Hamilton (BAH), a management consultant firm that previously had helped IHC

1978 International Model 86 Hydro. IHC offered the 86 Hydro with gas or diesel engines from 1976 into 1981. Gas models ran the 291-cubic-inch displacement in-line six while the diesels used the D-310.

reorganize engine production into the 300 and 400 series. BAH proposed IHC incorporate its foreign plants within the divisional structure in the United States. Brooks McCormick was the first to embrace this philosophy. European and Asian tractor operations all fell under Agricultural Equipment Group (AEG). Construction became the Payline Group (including smaller industrial machines). Truck Group and Solar Turbines International Group remained separate divisions.

As the farm equipment market improved, Brooks, who felt more confident after this latest reorganization, envisioned overtaking Deere as the farm equipment industry's number one manufacturer. Examining this goal forced McCormick to study the way IHC had recycled 1940s and 1950s technology during the 1960s and 1970s, which had expensive and sometimes

This gas-engine Hydro developed 69.6 horsepower off the PTO. It weighed 7,330 pounds and sold new for $17,165.

The company introduced hydrostatic drive 11 years earlier starting with the 656 in 1967. Over the following decade, hydro drive appeared on models ranging from the most powerful Model 186 down to Cub Cadets.

AL
86

1979 Model 3588. A fresh Iowa snowfall caught this 2+2 in the field. IHC introduced this double two-wheel-drive tractor concept to the marketplace in 1978 with this 150 horsepower version and a 130-horsepower Model 3388 at the same time.

disastrous results. The 86-series was a perfect example: Its engine was new, powerful and efficient but its frame, drivetrain, and controls were updated pieces introduced in the 706/806 models in 1963.

IHC watched Deere and Caterpillar produce big technological jumps in the 1960s and early 1970s. Experience and common sense told IHC engineers and board members that these two companies would need to leave these machines in production for 10 years to pay for them. This gave IHC room to introduce innovations in the 1980s that might allow them to move ahead and reclaim the title Deere had wrested away 20 years earlier.

1981 Model 7788. By the time this big tractor appeared, IHC had two variations of four-wheel drive tractors in its product line-up. This was a true articulated four-wheel-drive while the 2+2 blended the mechanics of two rear-drive models to create a similar function.

To ensure that plan continued on, McCormick handpicked Archie McCardell as his successor. McCardell joined IHC in late August 1977 as president under McCormick. Two years later, after McCormick stepped aside to make McCardell board chairman, McCardell tagged his personal choice, Warren T. Hayford, who left the aluminum can industry in mid-1979 to join IHC.

McCardell came from Xerox, where careful 5- and 10-year financial management and advance planning kept them in control of their market. However, strategic missteps left McCardell dissatisfied with Xerox and Xerox disappointed in him. McCormick knew IHC needed long-term planning and economic controls rather than short-term reaction funded by whatever resource was available instantly. Hayford brought with him ideas for cutting costs by increasing plant efficiency. Hayford's background, however, had not prepared him for the cyclical nature of farm equipment markets.

Inset
The Farm Equipment Research and Engineering Center, FEREC, chose its DVT-800 V-8 turbo-diesel to power these machines. With actual displacement of 798 cubic inches, these hard-workers churned out 265 horsepower at the PTO.

Below
Articulated four-wheel-drive tractor technology placed the operator's platform and cab above and ahead of the steering pivot point. IHC's 2+2 set the operator above the rear axle.

AGRICULTURAL EQUIPMENT GROUP SALES continued to climb, yet the Truck and the Payline Groups lagged behind. Inefficiency turned the two groups into IHC's loss leaders. McCardell increased the corporation's investment and research budget 25 percent in 1978 and 35 percent more in 1979. Money went into modernizing or building new plants, and the rest funded new products. He committed $150 million (about $525 million in 2005) to expand DT-466 engine production, and nearly $200 million (about $700 million in 2005) to another AEG project, dubbed TR-4 and TR-3A (the future 50-series and 30-series). To finance

IHC's engineers heavily reinforced these Model 6CH cabs. They were Roll-Over-Protection rated to 38,000 pounds while also incorporating air conditioning and a fully adjustable operator's seat.

This machine, the second of two produced, runs on eight 24.5R32 radial Firestone tires. North Dakota farmer/IHC collector Jay Graber considers this one of the jewels of his collection.

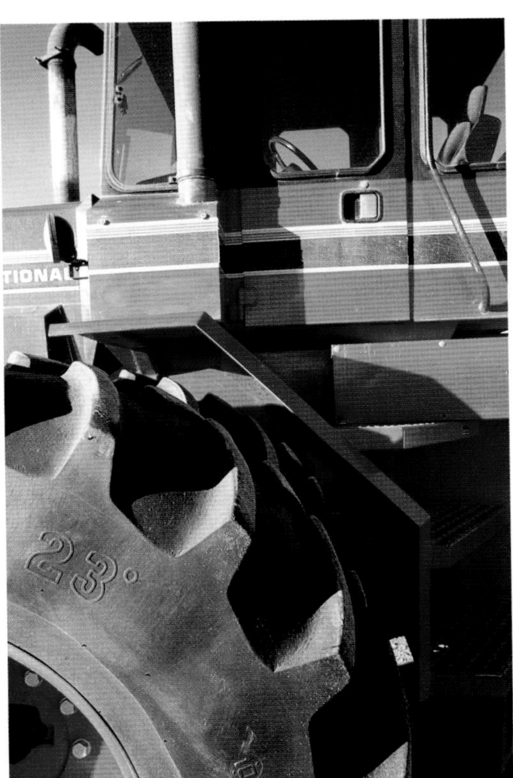

this, McCardell trimmed and clipped excesses and waste from everywhere within the corporation except the still-profitable Solar Turbines International Group. He managed to whittle $300 million ($1.05 billion in 2005) out of overhead and costs in 1978 and 1979.

Hayford took over McCardell's chores. He cut costs and introduced his can-industry work ethic, "the 8760 plan," representing the number of hours in the year. When Hayford arrived, IHC worked 14 of the 21 shifts a week, which, to his mind, meant one third of its capital resources were wasted in those 7 idle shifts. The number 8760 was a mantra to Hayford, but it became a battle cry to IHC's labor force.

There were three seasons and a distinct buying cycle for farm equipment. There was no season for aluminum cans. According to IHC historian Barbara Marsh, the first season occurred prior to spring planting when about a third of annual sales took place. The second opportunity was fall harvest, and this constituted half of annual totals because of combine and other high-priced equipment sales. The last appeared in late December for year-end tax planning. This cycle, as it had evolved since the 1950s, produced one peak sales year in every seven. IHC saw a peak in 1966 and again in 1973, so it was looking forward to 1980. It was pleasantly surprised to get the peak in 1979. But, that was the year that nearly every good thing in the short term turned out to be bad in the long run.

On May 19, 1977, the newly named Agricultural Equipment Group (AEG) approved an earlier FED development and testing request. That involved powertrains for a series of larger-horsepower tractors, along with

This was the view from the driver's seat of the 23,800-pound, 20-forward-speed 7788. IHC assembled just two of these, and sold them new for $93,320.

VOLTS
FAST
SLOW

Introduced in 1983, these 2-wheel-drive tractors weighed 11,542 pounds. They sold new for $49,840.

1984 Model 5488. These were among the last models to wear the International name. The company produced these in two- and four-wheel-drive versions.

Matt Jackson's working tractor runs on dual 20.8R38 rear radials. Using IHC's DT-466 in-line six-cylinder diesel, this machine produced 187.2 horsepower at the PTO, good reason to need dual rubber to get the power to the ground.

Both two- and four-wheel-drive versions used IHC's D-466 in-line six with 112.5 PTO horsepower.

a new pressure-flow-compensating hydraulic system, as well as a combined project originally referred to as "MATH," for "Modular Axle, Transmissions, Hydraulics," that IHC had first used in CED's TD-20E crawler.

The Modular Axle, described in a 1978 engineering data report, was a "final drive housing with a new cast center housing containing an increased capacity differential with spiral bevel reduction and inboard-mounted planetary final drive . . . . An integral differential lock electrically energized will be offered as an option.

"In addition to being the final drive on two-wheel drive tractors, this assembly in modular axle form will be used as a front and rear unit on large four-wheel drive articulated tractors."

The pressure-flow-compensated pump provided low hydraulic pressure for steering, transmission, hydraulic-oil cooling system, independent PTO, and wet multiple disc brakes. The

1984 Model 3488 Hydro. This was the model IHC engineers spent time developing and this was the model the company hoped would help save it. Introduced in 1981, the 88-series represented the best engineering-into-the-future, with marvelous transmissions and powerful engines.

high-pressure side took care of lubrication, draft control, and auxiliary valves. While the highest-proposed output for future engine versions of the proposed 50-series, known internally as the TR-4, was 185 horsepower, AEG rated the transmission at 200 horsepower, needing only to revise PTO clutches to accommodate even more power. FED planned to produce three models. One would rate 135-PTO horsepower; the second would develop 160 horsepower; and the largest had 185 horsepower using Melrose Park's new DTI-466 turbocharged and intercooled diesel. "The tractor front end will be restyled for model identification and to conform to new tractor family lines," the report continued.

The greatest technical advancements were in transmissions offered with the 50-series TR-4 and the lower-horsepower-range 30-series, known inside AEG as the TR-3A tractors. For the

The standard-equipment Synchrotorque TR4 transmission provided 18 forward speeds with full synchromesh shift-on-the-go capability. With single rear tires, the tractor weighed 14,061 pounds.

3A series, FED had in mind the new P-3A Constant Mesh (synchromesh) transmission. For TR-4 models, FED was completing either a Synchro-Torque or a Vari-Range transmission. Engineering planned to introduce these transmissions in the 2+2 models as well. The Synchro-Torque transmission required a clean lab "white room" to assemble the internal parts. This incorporated a three-speed constant-mesh gearbox, a two-speed double-clutch-pack Torque-Amplifier unit, and a three-range constant-mesh set (3x2x3) providing 18 forward and six reverse speeds. FED also planned to produce a high-clearance version of the TR-3A. Louisville and Farmall Works would assemble the tractors, with production to start in October 1984. FEREC's report suggested that: "The program proposed in this request is the basic building block for development of high horsepower agricultural tractors well into the 1990s. The hydro mechanical transmission becomes the foundation for further refinement of tractors through development of electronic monitoring/control of critical functions and, conceivably, fully automated tractor operation." AEG now had taken hold of the technology of the late 1970s and was looking far ahead.

5488
INTERNATIONAL

The two-wheel-drive models sold new for $66,515. IHC priced the four-wheel-drive versions at $78,340.

1985 Model 7288 2+2. Jeff Kelich put his 2+2 to work pulling his Model 720 6-18 On-Land plow through hard dry soil in Indiana. Spring shock mounts kicked the plow bottoms out of the soil upon impact with a rock.

The On-Land plow allowed the tractor to remain flat and level on the unplowed land to the left of the furrows. The 18-inch bottoms dug deep into the dry soil.

This was the operator's view. Turning the wheel of the 2+2 swiveled the long nose ahead of the platform and brought the cab and rear axle around to follow it.

However, Engineering needed time with the Synchro-Torque system and the Vari-Range transmissions to perfect the compound planetaries, and Manufacturing was moving slowly due to the major tooling necessary. R. J. Roman, the TR-4 project manager, alerted J. T. Tracy, AEG's director of Product Planning and Development that, "As a result, plans to incorporate these two new transmissions on 88-series 2+2s would be delayed until May 1982." H. B. Simmons was not happy. As manager of the financial responsibility arm of AEG, he knew a year's delay was lethal to planning, marketing, and sales.

While internal squabbles in late 1979 dealt with the products designated to restore IHC to first place, factors outside AEG were conspiring to make that much more difficult.

BY YEAR'S END, SALES FROM ALL GROUPS had reached record levels. McCardell's economics and Hayford's cost cutting had provided the corporation its highest profit margin in 10 years. McCardell restructured

some of IHC's debt into short-term financing, terminating at the end of 1981, rather than relying on the typical long-term arrangements used for such expenses as new factories and tooling.

Manufacturing in all groups increased production through the spring and summer in advance of labor union contract negotiations and a strike by UAW (United Auto Workers) workers that IHC management expected. According to historian Barbara Marsh, IHC added "$125 million of additional inventory to sell in case the strike went on any period of time" (equal to about $562.5 million in 2005 dollars). Because of higher profits, greater sales, raised production levels, and several other favorable factors, a complex formula fell into place that came together at the worst possible time to reward McCardell with a $1.8 million bonus ($6.3 million in 2005). This was not a good prelude to requests that workers and the union make sacrifices when talks began August 9, 1979.

The UAW walked out on November 2, the day after IHC published annual figures. The 35,000 strikers received $50 ($175 in 2005 dollars) per week. It hardened their resolve against IHC, Archie McCardell, and his bonus. While McCardell ultimately found a legal way to return the bonus, temporarily, too much damage had been done. Then it got worse. Neither he nor Brooks McCormick wanted a long strike, despite their high inventory

IHC produced the 7288 in 1985 only and this is the 4th of just 19. Powered by IHC's DT-466 turbocharged diesel, the tractor boasted 175 PTO horsepower.

The 2+2 designation came from the concept of combining two two-wheel-drive tractors together. This was a new answer to the question of how to get more power and traction to the ground.

Most commonly seen with dual wheels all around, this 7288 arrived for an IH club Plow Day event without needing an oversize load permit. In this configuration it weighed 19,750 pounds.

preparations. Yet neither man would interfere in the negotiations, trusting a succession of ill-prepared but unyielding company representatives.

The strike lingered. IHC's long- and short-term debt went unserviced. Sales in all three groups dwindled after the Federal Reserve Bank raised an inflation-fed prime rate to 20 percent, and consumer rates flew higher still. In March, IHC realized that what labor demanded was less threatening than its financial crisis. Working at the bargaining tables almost without break, they settled on April 14, 1980, six months after the strike had begun. The strike had cost $579.4 million ($2.03 billion, adjusted to 2005 values) in losses during the first half of fiscal 1980 (November 1 to October 31).

Without the plow and without its typical dual wheels, IHC quoted the list price for this tractor at $73,500. Most buyers saw deep discounts.

STS
INTERNATIONAL
7288

1985 International Model 7488. IHC assembled most all of these "Super 70s" by hand at the Farmall Works during January of 1985. By sunset, May 14, all tractor manufacture had ended.

No one at IHC could have predicted the future. Had the strike not occurred, would IHC still be in business? Maybe, but there were other pitfalls awaiting the corporation. At the very least, IHC would have entered 1980 and its shaky economy with an enormous inventory that was growing by the day, fed by a man fixated on a number: 8760.

Hayford's 8760 philosophy came with very high overhead costs. With all plants working 21 shifts per week, there could be hundreds of expensive assembly-line workers standing around if major tooling failed. Production would stop. In Hayford's perfect world workers changed shifts in the instant between tasks when their replacement stepped in and picked up their tools. Real-world assemblers don't perform with the rehearsed precision of relay-team baton handoffs.

Hayford had proposed that critical plants such as Farmall, Louisville, and the converted diesel plant at Indianapolis duplicate tooling and partially replicate assembly lines. While one was in use, the other would be there "just in case." Lack of available funds made this impossible. At

Farmall Works, supervisors had developed cheaper techniques for handling tooling breakdowns. They hauled out pieces that needed machining to friendly department heads at the Rock Island Arsenal, Caterpillar, or even Deere, while IHC's plant crews repaired the breakdown. When the shoe was on the other foot and IHC's machinists had time, they returned the favor.

The market for all IHC's products had slowed. Yet Hayford saw only inefficient, idled plants that, in his view, cost the company money when they weren't turning out products that made the company money; that nobody was buying didn't matter. Fiscal 1980 ended with tractor sales 14 percent lower than 1979. Overall sales dropped by 29 percent. McCardell reduced his ambitious capital-spending plan for 1981 by one third, scaling back plant construction, acquisition, tooling, and new-product development.

McCardell, despite his financial background, believed IHC's problems could all be solved if the company just had a couple of great selling seasons. Hayford agreed. They fed each other's faith in a future for IHC. Groups would need new products. Funding for the TR-4 and TR-3A programs continued.

Its internal code designation was TX-194. With 200 PTO horsepower, IHC priced the tractor at $78,675.

1990 Case-IH Maxxum Model 7140. Case-IH introduced these models in 1987 and they remained in production into 1994. The two-wheel drive models appeared in 1988. Their Case-IH six-cylinder 504.5-cubic-inch turbo diesel developed 197.5 horsepower on the PTO.

The four-wheel drive models weighed 16,578 pounds with single rear wheels fitted. List price was $103,180, about $12,000 more than the two-wheel drive version.

In January 1980, 2+2 project manager R. J. Roman went back to AEG president Bill Warren to ask him for more money. A synchro-shift mechanism from Clark Equipment Company did not prove strong enough. A worldwide search for alternatives found nothing useful, so FEREC developed their own. It cost $8 million ($28 million in 2005 dollars). Revising the air-flow cooling system and relocating operating controls to the Control Center right side cost $1 million each ($3.5 million in 2005). Other component design and development consumed another $8 million ($28 million), while production success of the 2+2s from Farmall Works meant manufacturing facilities intended for TR-4 use were not available. Farmall Works started to construct a plant addition.

Development of the Synchro-Torque transmission resulted in improvements, simplifications, and standardizations that would benefit the Vari-Range. Another $4.2 million ($14.7 million in 2005 dollars) would apply these updates to the state-of-the-art transmission. Roman allocated $5.8 million ($20.3 million) to accommodate the effects of 1980s inflation, adding a

The two-wheel drive models weighed 15,630 pounds. The mechanical front-wheel-drive hardware added another 945 pounds. Both versions offered 155 horsepower at the PTO.

total of $34.4 million ($120.4 million) to the $193 million project ($665 million, adjusted for 2005 inflation).

The success of the 2+2s accelerated development of a 195-PTO-horsepower version. The drivetrain was marginal above 180 horsepower, so engineering now considered the 195 model as a segment of the Synchro-Torque TR-4 program that became the 7288 and 7488 2+2 "Super 70" models. Complete sign-off engineering prototypes were not scheduled until August 1980, but even that extended date was delayed when problems appeared in late spring.

Meanwhile, failures during final testing of the 3788 tractor with the old 86-series transmission, suggested final-drive reliability problems would occur in the second or third year of average operation, or just beyond the warranty period. The 3788 powertrain was a risk in terms of future farm tractor liability. These were not small problems. Engineers at the product reliability

1998 Case-IH Magnum Model 8920. Case manufactured this series in two-wheel drive and fitted with mechanical front-wheel drive models. Production spanned 1997 and 1998. The 6-cylinder turbo diesel displaced 505 cubic inches.

and support center near Hinsdale suggested these types of failures may require a rebuild of both the range and speed transmissions. Taking Marketing's estimates of three-year sales at 1,695 units, they calculated 55.5 hours per repair at $18 per hour ($63 in 2005 dollars) shop costs, or $1,000 labor ($3,500), plus $2,300 in parts ($8,050) at dealer net, or $3,300 ($11,550) per tractor. The total: $5.5 million ($19.25 million in 2005 dollars). "With the high load factor projected on the 3788 tractors, the failure cost is on the conservative side."

AEG scheduled production to start August 13, 1980. A frenzy of meetings followed in which Marketing worried over releasing a transmission that AEG knew would fail. Engineers fired back a detailed analysis of its revisions and fixes. Instead of delaying it, AEG advanced the start of production to August 5.

Two months later, on October 7, Marketing approached H. B. Simmons from AEG's Business unit. Marketing was concerned about large four-wheel-drive sales. IHC's percentage of the business had fallen steadily since 1976, when it had reached 14.4 percent of the total market. By 1979 it was only 10.8 percent, and Marketing projected it at 8 percent for 1980. They feared that even with 1981 improvements to the Steiger articulated 4WDs, IHC still had no powershift or PTO option for those tractors, although they had been proposed. Simmons agreed, penning across the top of Marketing's memo:

"If we don't get the changes for '81 we must have—we will be out of this business." Simmons meant the articulated 4WD business. However, his words were prophetic.

Case-IH 1998 Maxxum Model MX-120. Case kept these tractors in production from 1996 into 2002. The company offered them with mechanical front-wheel drive as shown here as well as two-wheel drive.

Case provided these tractors with the 18-forward-speed transmissions. These tractors sold new for around $85,000.

In late October 1980, the TR-4 and TR-3A programs asked for another $9.4 million ($32.9 million in 2005 dollars). Of this, $6.5 million ($22.75 million) would make up engineering time lost during the UAW strike, which occurred during peak development and testing for the new tractors. Now, AEG reassigned more than 100 design engineers and managers who had shop skills to move the project forward. The remaining $2.9 million ($10.2 million) was for design, assembly, and testing of additional prototypes, since some elements done during the strike failed and needed redesign. Assembly of the TR-4 involved 220 new major machine tools requiring new automated and special transfer line equipment. A wary Manufacturing Department planned to build 100 preproduction tractors in May 1981.

In February 1981, Marketing's J. W. "Bud" Youle drove prototype TR-3A and TR-4 tractors at the Phoenix Proving Grounds. He was impressed and made that clear in his memo to Simmons and project manager Roman. "Without question, the TR-3A was needed ten years ago. It is unquestionably the easiest shifting tractor that I have ever driven . . . . In the writer's opinion, there is no doubt that the TR-4 is 'TOMORROW'S TRACTOR TODAY,'" he wrote, mimicking an advertising slogan for IHC's 2+2s.

The situation inside IHC was approaching critical mass. In May 1981, to stanch the outward bleeding of cash, McCardell had "sacrificed" IHC's only profitable asset, the Solar Turbines International Group, to Caterpillar for $505 million ($1.77 billion in 2005 dollars). The unexpectedly high price helped briefly. By early spring 1982, McCardell proposed yet another restructuring of management and debt. Hayford suggested taking the corporation into Chapter 11 bankruptcy. In March, fed up and without hope, Hayford resigned.

Marketing had its own problems. It had to suffer seeing a prediction come true in late April 1982, on the eve of Engineering's release to Manufacturing of the 6788 that would replace the 3788.

"Field inspections by Product Reliability," Marketing wrote in a memo to World Headquarters, have "confirmed original concerns regarding the 3788 power train in reliability and hours of usage . . . . It is the writer's opinion the 6788 will not meet or, much less, exceed, the 3788 reliability of the standards now established by the Series 50 [TR-4] power train. There should be no misunderstanding with management of the reliability risk with the planned production of the 6788."

Inside this optional cab, the operator had a choice of 16 forward and 12 reverse speeds. The 359 cubic-inch displacement in-line six-cylinder turbo diesel developed 105 horsepower at the PTO.

The tractors came out: the TR-3A, 30-series in August 1981, the 50-series and revised 2+2s just reaching dealers. The economy that hamstrung IHC also hurt farmers. Federal bans on Soviet grain sales in 1980 gave foreign farmers a windfall market but left their own countries hungry. American farmers who no longer fed the communists nourished others briefly. But President Ronald Reagan's fiscal policies elevated the value of the dollar so greatly that few countries could afford American produce. By 1982, American farmers were becoming an endangered species. So were healthy farm equipment dealers.

Huge stocks of IHC's previous models remained on distributors' and dealers' lots. When the 30- and 50-series arrived, churned out by plants running on the theory that large profits came from volume sales, they arrived at dealerships where few people were interested in, or capable of, buying them. Payment-In-Kind (PIK), a federal program to reduce crop acreages and surpluses, dealt farmers another severe blow when the 1983 drought reduced harvests in the fall. Foreign farmers hurried to feed Americans, accepting strong dollars in payment.

On August 10, 1983, Donald D. Lennox, the new president and chief executive officer, replacing Warren Hayford, signed off on the Super 70s, the 7288, 7488, and 7788 models, with little additional development in response to Marketing's memo. Lennox authorized a $16 million (approximately $48 million in 2005 dollars) engineering completion budget, an August 1984 start, and a $46 million ($138 million) working capital commitment through 1987. Lennox and new AEG president J. D. Michaels released the Super 70s as well as the 175-horsepower 7488 model and the 195-horsepower 7688. Each of these offered full 50-series features including the

Case-IH 1998 Quad-Trac Model 9380. These were go-anywhere machines. Case manufactured this model from 1996 into 2001.

Vari-Range transmission, a "Fuel-Efficient" engine management program, a 40-gallon-per-minute high-pressure hydraulic pump, and the new Control Center cab.

The rear half of the 70-series 2+2s was very similar to the rear of the 50-series 2WD tractors, and it was manufactured on the same line. Shop staff then transferred that half out to a new 78,000-square-foot building completed in 1978, for $18.9 million ($66.2 million in 2005). In this structure, dedicated to assembling and painting 2+2s, and manufacturing the front of the 70-series tractors, workers joined the halves and completed final assembly.

Part of the launch program, scheduled June 1984 through April 1985, included damage control. From October through December 1984, IHC held customer and dealer Weather Vane meetings, "to improve the customer image of the 2+2 concept and to provide them with detailed changes as to why the new 2+2 will be more reliable in the future."

BY JANUARY 1984, THE FEELING OF CONCERN spread far beyond North Michigan Avenue. Marketing wrote another memo late that month to AEG product marketing manager, A. W. Williams, pointing out shortcomings of 50-Series tractors and Control Center cabs with door and control lever placements. Due to drastically reduced capital spending and development budgets, corrections and improvements had to be slipped back to 1987 for introduction.

"I'm aware," a Marketing executive wrote, "of the numerous restraints you have concerning budgets, etc. However, we cannot wait until the 1987 sales year to correct dealer and customer objection to our Farmall tractors."

It was far too late. Don Lennox had launched a strategy to save the corporation in 1981. After its success that year, it gathered momentum. Tractor Equipment Company, a major parts supplier, heard that Dresser Industries, Inc., a Texas-based oil exploration and development conglomerate, was looking to add construction equipment to their line. IHC's Payline Group, with plenty of inventory on hand, fit their profile. Dresser struck a deal, acquiring the Hough operations as well as IHC Payline later that year. This left Don Lennox one less group to focus on. Through 1984, he conferred with bankers and courted a potential buyer. On November 26, 1984, following hundreds of hours of work and negotiation, IHC agreed to sell AEG to Tenneco, Inc. of Houston, another large conglomerate with primary interests in oil production but with a wholly owned agricultural subsidiary, J.I. Case. IHC received $301 million in cash (about $900 million in 2005 dollars), and $187 million ($561 million) in Tenneco preferred stock.

Case acquired engines from Cummins. This was the N-14 in-line six-cylinder turbo diesel with 855 cubic inches of displacement. Cummins and Case rated it at 355 horsepower at the PTO.

Case manufactured about 20 farm tractors. It had quit the harvester business long before, but it wanted IHC's axial-flow combines. Tenneco did not want IHC's Rock Island or Memphis plants. Overnight, Tenneco and IHC increased Case's market share to 35 percent, making it a contented runner-up to Deere, which had 40 percent. IHC brought 33 tractor models to the sale. The new company, Case IH, dropped the 2+2 and new 30- and 50-series tractors because they didn't want Farmall Works, and they carefully selected the tooling they wished to own.

On May 14, 1985, the last International tractor, a Model 5488 All-Wheel Drive, came off the assembly line at Farmall Works. The Payline Group had moved to Texas, and Trucks remained in Chicago as Navistar. The Cub Cadet was safe with MTD Corporation in Cleveland, Ohio. But one of America's great legends was gone.

With 12 speeds forward and two available for reverse, these machines often found use on road construction sites. The big Quad-Trac weighed 43,000 pounds.

In 1983, due to an EPA requirement, Case had abandoned its Power Red and Power White paint scheme because there was too much lead chromate in its paints. It adopted a white-and-black combination until early in 1985, when sheet metal became Harvester red while the chassis remained Case black. Nearly all of IHC's popular Neuss-built and Doncaster-produced tractors remained in production, providing Case IH much greater European name recognition than before. Case's own 94-series tractors, built at its Racine, Wisconsin, plant, replaced all the domestic-built AEG models.

IHC's engineers had nearly completed work on their powershift transmission. They mated this with Case chassis, engines, and bodywork. This combination, introduced by Case IH in 1988 and kept in production until late 1993, became the Magnum line of 130- to 195-PTO-horsepower two- and four-wheel-drive tractors. In September of that year Case IH introduced the Second Generation Magnums.

Prior to acquiring IHC, Case engineers had begun work on the Maxxum series, Case's World Tractor, offered with an adjustable front axle in two-wheel-drive or mechanical front-wheel-drive assist. IHC engineers joined Case early in that process. When Engineering completed its prototypes, they went to IHC's FEREC in Hinsdale for testing and customer marketing trials. After completing engineering work, the former IHC Neuss Works did the styling and assembly. Case IH introduced its 5100-series in 1989 and the Second Generation 5200s arrived in late 1992.

In 1987, Tenneco had acquired Steiger Tractor Company, of Fargo, North Dakota, which had previously produced articulated 4WD models for IHC. Steiger was a victim of the early 1980s farm crisis and had gone into Chapter 11 bankruptcy. Quickly re-badged, Case IH introduced its 9100-series still using Steiger model names: Puma, Cougar, Tiger, and others in 1987. In August 1990, the Second Generation Case IH 9200-series of Fargo-built 4WDs appeared. Nearly a year earlier, Steiger had begun field-testing in North and South Dakota a multitrack prototype based on the 9250. Tests went on carefully for three years, but few people saw the machine because operators ran it only in the dark where, from a distance, it was indistinguishable from the wheeled model. A shiny version toured Farm Progress and other major shows during 1992, but it was billed more as a "concept tractor" than a work-in-progress. Case IH meant to show Caterpillar that it was not alone in thinking about rubber tracks.

With its retail price set at $211,791, the Quad-Trac offered serious power and versatility for the operator who needed it. Its articulated steering and high drive sometimes allowed these machines to pull other crawlers out of the muck.

Nighttime tractor development continued, but Case IH also tested the technology on harvesters, especially in California's rice country north of Sacramento in late summer 1993. When they introduced the production QuadTrac to dealers in Denver in 1996, those who'd seen the show model were surprised at the growth. It had gone from 246 horsepower to the 360-horsepower 9370 chassis. In mid-1998 Case IH introduced a second model, a 400-horsepower 9380. Model proliferation continued. The horsepower race never ended.

The New Millenium Brings a New Identity and the Return of the Farmall

Market garden (or "truck" to use the ancient terminology) farmers demand versatile, tough, and small equipment for their operations, which don't stop for bad weather or bad operators. This DX29 is at work in pleasant conditions in Pennsylvania. *Photo courtesy of CNH America LLC*

Chapter 11

2000 and Beyond

"We see our customers using larger implements with higher horsepower and hydraulic requirements," Stacey Smith said in an interview conducted for Case IH's Media Department. Smith was the company's marketing and training manager for the company's larger tractors, those with more than 100 horsepower. In mid-July 2002, the company launched the new MX-series Magnum row crops with 505-ci displacement, six-cylinder engines ranging from 170- up to 240-PTO horsepower. These were technological marvels, with computers and electronics that communicated between the tractor, the implement, and global-positioning satellite (GPS) systems to precisely control tractor speed, direction, and implement work meter by meter. Front dual tires, available across the entire lineup since 1998, now rivaled Case's revolutionary QuadTrac for flotation and low soil compaction. A new luxury cab offered operators exceptional comfort with the Positive Response seat. This seat interfaced with a sensor and controller to adjust the shock-absorber cushioning impact to the cab as often as 500 times per second. MX cabs also used a soy product to fabricate the roofs.

CASE
LX114

"Our customers help us design equipment that fits their needs," Drew Fletcher explained in a release published by Case IH Public Relations. Fletcher was product specialist for Case IH tractors. "Obviously, the amount of ground being worked by individual operators and new tillage practices require more horsepower today to get the work done." Case IH celebrated a milestone on November 25, 2003 after assembling its 100,000th Maxxum since introducing the line in 1987.

At the same time that Case IH brought out the big MX-series, it also introduced a midrange series, the MXM Maxxums. It offered these models on two different wheelbases. Their 456-ci-displacement, six-cylinder diesels ranged in output from 95- to 160-PTO horsepower in either two-wheel-drive or mechanical front-wheel-drive configurations. New suspended front axles provided 4.2 inches of vertical movement without sacrificing traction, steering control, or stability. Electronics allowed operators to record as many as 28 tasks, programming into their tractor the operations it must accomplish at the end of each row, from raising the implement or shifting gears, to changing engine speeds, and more.

The Farmall may have been responsible for replacing more horses than any other machine in history. Now horses are recreational, and the Farmall cares for them on this Georgia layout. The DX29 has 29 horses of its own under the hood. *Photo courtesy of CNH America LLC*

Today's modern Farmalls come with FOPS folding rollover protection, which allows the tractor to fit into low overhead buildings, while still giving the operator rollover protection outside. This is critical in loader operations. *Photo courtesy of CNH America LLC*

These new tractors were all part of the extensive product lineup created when Case IH merged its forces with New Holland in late 1999 and became CNH. "On November 11, 1999, Fiat purchased the outstanding shares of Case Corporation stock for $55 per share in a friendly takeover." The resulting operation became the number-one manufacturer of agricultural tractors and combines in the world. A reorganization within CNH, the parent company, in January 2003, divided Manufacturing, Marketing, and financial responsibilities among three large geographic regions, "moving operations closer to the customer," Paolo Monferino

Anyone who has raised horses knows they're a heck of a lot of work, which is one reason why the Farmall was so popular in the first place. This DX29 with matched LX114 loader and weight box is making the work at least easier! *Photo courtesy of CNH America LLC*

One new twist to the new Farmall line is the availability of matched rototillers for the large gardener or truck farmer. This DX29 is powering a Case-IH TLX180 rotary tiller to produce a pulverized seedbed. The LX114 loader probably stays on full-time. *Photo courtesy of CNH America LLC*

explained in interviews with Jeff Walsh, Corporate Communications director. Monferino, CNH president and CEO, added that this change enabled CNH "to better respond to customer needs." In this new organization, product design and development were to remain part of a global decision process with local influence. Rich Christman directed agricultural business activities throughout North America, Australia, and New Zealand. Christman was a former design engineer with Ford Motor Company; he joined Case in 1975 and Case IH's board elected him president of its agricultural business operations in July 2000.

On January 22, 2004, CNH and its Case IH agricultural operations announced the resurrection of a legend. With a power range roughly equivalent to what IHC's Farmall offered throughout its first 40 years, Case IH reintroduced the Farmall name on a new product line directed at buyers ranging from hobby farmers up to full-time farmers, landscapers, or commercial contractors. The new machines, in D- and DX-series, stretched from 13.7- to 47-PTO horsepower with gear or hydrostatic transmissions and many of them with standard MFWD. Bill Frederick, Case IH's Product Marketing and Training manager for compact tractors, explained the philosophy behind the new lineup in materials released at the time of the announcement.

Often forgotten in the history of the Farmall is that thousands of them ended up in mowing jobs on golf courses and roadsides from the start of regular production in 1924 on. The DX29 is available with either a 9x3 gear transmission or, in keeping with IH tradition, a Hydrostatic. All have mechanical front wheel assist. *Photo courtesy of CNH America LLC*

"Most new tractor purchases are driven by the subcompact and utility market. It represents seventy-two percent of all tractors sold in North America. While operator comfort is a priority for the experienced operator," Frederick continued, "it's even more critical for the inexperienced or first-time user, since approximately seventy-five percent of these units are purchased for non-farm applications." With full hydraulics to operate a front loader or backhoe, and a three-point hitch on which to mount anything from an aerator to a tiller, the new Farmall also boasts easily accessible controls and a swiveling seat.

As of late 2004, CNH, the parent company, comprised a network of dealers and distributors across 160 countries throughout North America, Europe, Asia, the Middle East, Africa, Australia, New Zealand, and Latin America, selling Case IH, New Holland, and Steyr agricultural products.

While the current crop of Farmalls are small tractors, they can go over a windrow while working with a round baler. The DX55 pictured opposite is the largest in the current Case-IH Farmall line. *Photo courtesy of CNH America LLC*

IHC'S ORIGINAL FARMALL TRACTOR LINE was a hard act to follow. While the tricycle-configuration tractor didn't defeat Henry Ford, it did humble and redirect Deere & Company's general-purpose tractor-design efforts after it appeared. For decades, up through the 1970s, other manufacturers followed IHC's lead, and the tricycle configuration was the definition of mechanized row-crop farming. Throughout that 50-year period, however, engineering improved and safety became an increasing concern. Tricycles were inherently unstable unless the operator had nearly perfect conditions. The style and concept of standard-configuration machines such as Ed Johnston's International 8-16 ultimately proved more enduring. Even the influence of Johnston's slope-away front hood was more apparent by the mid-1980s and remained so into the new millennium.

In its 82-year life, IHC's greater- and lesser-known farm tractors advanced the benchmark many times. Ed Johnston's huge Moguls and Harry Waterman's Titans opened the Great Plains and Canada, and Tractor Works' gear-driven 10-20 and 15-30 McCormick-Deering tractors challenged draft animals and other competitors, as well. Bert Benjamin redefined mechanized

The DX55 has, depending on transmission, either 46 or 47 PTO horsepower. Between matched mounted implements, pull-behind units, and the great commonality of the three point hitch, the Farmall can be used with a large variety of equipment. *Photo courtesy of CNH America LLC*

farming with his Farmall, and added a word to the vocabulary. In parts of the Midwest and south, as recently as the early 1990s, some people still called any tractor a Farmall, whether it was gray, red, green, blue, yellow, orange, or silver.

The F-series and its counterpart W-series four-wheel tractors took additional hundreds of farmers away from mules and horses, helping feed the United States and the world during the 1940s. But the war distracted International Harvester Corporation and convinced its board that it could be much more than merely an agricultural equipment and truck manufacturer. Fowler McCormick and John McCaffrey, hoping to move IHC into an even greater manufacturing prominence for the last half of the twentieth century, became distracted by the possibilities. An up-and-down economy was no help in keeping their focus on what IHC should do. Hundred-series tractors left blemishes on IHC's reputation as did its biggest crawlers. But the thousand-series agricultural tractors and its Hough (and later Steiger) collaborations restored the name International Harvester. Sales competed with Engineering on product ideas and delivery dates, often casting Engineering into the subservient role.

Today's haymakers have a choice of bale sizes and shapes. This one's using a DX55 with a bale spear to handle a small round bale with the three point lift system. The labor reduction from small square bales is immense! *Photo courtesy of CNH America LLC*

There was no single, simple cause to report for the end of IHC, and there is no single individual, after nearly 20 years, on whom to lay the blame. It was the result of nearly 40 years of subtle miscues and slight missteps. IHC was a generous employer and a manufacturer that earned lifelong loyalty from its customers. It remains a matter of fierce pride to red-tractor owners when they speak with friends who own other makes. That pride of ownership, of operation, of performance, of tractor quality, is why Case IH has brought the Farmall name back. Time will tell if the new D- and DX-series tractors will carry on the tradition and reestablish the legacy, or if they will invent a new significance for the name Farmall. It will be the job of subsequent historians to recount the legend of the new Farmall.

INDEX